Finding Mutations

The Basics

Series editors: R. J. Beynon, T. A. Brown, and C. J. Howe
Series advisors: T. Hunt and C. F. Higgins

Finding Mutations
The Basics

J. R. Hawkins

*Department of Paediatrics,
Cambridge University
Addenbrooke's Hospital,
Cambridge*

◯ **IRL PRESS**
——at——
OXFORD UNIVERSITY PRESS
Oxford New York Tokyo

1997

Oxford University Press, Great Clarendon Street, Oxford OX2 6DP

Oxford   New York
Athens   Auckland   Bangkok   Bogota   Bombay   Buenos Aires
Calcutta   Cape Town   Dar es Salaam   Delhi   Florence   Hong Kong
Istanbul   Karachi   Kuala Lumpur   Madras   Madrid   Melbourne
Mexico City   Nairobi   Paris   Singapore   Taipei   Tokyo   Toronto

and associated companies in
Berlin   Ibadan

Oxford is a trade mark of Oxford University Press

Published in the United States
by Oxford University Press Inc., New York

© Oxford University Press, 1997

A catalogue record for this book is available from the British Library

Library of Congress Cataloging in Publication Data
(Data applied for)

ISBN 0 19 963611 7

Typeset by EXPO Holdings, Malaysia

Printed in Great Britain by Bath Press Ltd., Bath

Preface

The ability to detect and characterize mutations is fundamental to an understanding of biological function. In these days of genome projects and DNA sequencing on a massive scale, mutation detection is a major rate limiting step in understanding the function of particular genes.

The great majority of mutation detection is currently carried out on human patient material. As a consequence, this book has a definite slant towards Medical Genetics. The same techniques of mutation detection are, however, equally applicable to mice, flies, fish, or plants.

Despite the importance of mutation detection in today's biology, there is not one really good way of doing it. Instead, there are a variety of different methods, each with their own intricacies, merits, and drawbacks.

The aim of this book is to show what each method is capable of, how it works, and briefly, how it is performed. The book is definitely not intended to be a laboratory protocol handbook, but is intended to give the uninitiated and the novice (and maybe even the experienced too) an understanding of the methods available and which is most appropriate for the problem at hand.

The methods section of the book is divided into two chapters: one on diagnostic mutation-detection methods—those which determine the presence or absence of a particular, previously characterized mutation; and the other chapter on scanning mutation–detection methods—those which detect unknown mutations.

The fear in writing a book on molecular biology methods is that since the field is fast moving, the book will be out of date by the time of publication. In a bid to anticipate any significant developments, the final chapter provides a survey of new but unproven technologies which might become important. It will be up to you, the reader, to monitor any new developments in these technologies.

I am grateful to many members of the Cambridge University Genetics Department, especially Alan Schafer, Patrick Hextall, Ricky Critcher, Simon Whitfield, and Christine Farr for answering the most basic and complex questions; to Sarah Freestone, Andrew Palin, and Rob Beynor for criticisms and suggestions; and to Peter Goodfellow and Ieuan Hughes for tolerating time lost to the book.

J. R. H.

Cambridge
March 1997

Contents

Abbreviations

A	2′-deoxyadenosine (in a DNA sequence)
AP	alkaline phosphatase
ATP	adenosine 5′-triphosphate
bis	*N,N′*-methylenebisacrylamide
bp	base pair
C	2′-deoxycytidine (in a DNA sequence)
cDNA	complementary DNA
Ci	Curie
CpG	C and G dinucleotide pair (running 5′ to 3′)
ddATP	2′,3′-dideoxyadenosine 5′-triphosphate
ddCTP	2′,3′-dideoxycytidine 5′-triphosphate
ddGTP	2′,3′-dideoxyguanosine 5′-triphosphate
ddNTP	2′,3′-dideoxynucleoside 5′-triphosphate
ddTTP	2′,3′-dideoxythymidine 5′-triphosphate
DEPC	diethylpyrocarbonate
dGTP	2′-deoxyguanosine 5′-triphosphate
DMSO	dimethylsulphoxide
DNA	deoxyribonucleic acid
DNase	deoxyribonuclease
dNTP	2′-deoxynucleoside 5′-triphosphate
dsDNA	double-stranded DNA
DTT	dithiothreitol
EDTA	ethylenediaminetetra-acetic acid
ELISA	enzyme-linked immunosorbent assay
e-mail	electronic mail
FISH	fluorescent *in-situ* hybridization
G	2′-deoxyguanosine (in a DNA sequence)
HRP	horseradish peroxidase
kb	kilobase pair
kDa	kiloDalton
Mb	million base pairs
mRNA	messenger RNA
PCR	polymerase chain reaction
RNA	ribonucleic acid
RNase	ribonuclease
SDS	sodium dodecyl sulphate
SSC	sodium salt citrate
ssDNA	single-stranded DNA
T	2′-deoxythymidine (in a DNA sequence)
Taq	*Thermus aquaticus*
TEMED	*N,N,N′,N′*-tetramethylethylenediamine
T_m	melting temperature
UV	ultraviolet

Introduction to mutation detection

1. Introduction to mutations and their detection

Mutation is an alteration in an individual organism's genetic material. It occurs as a spontaneous or mutagen-induced event and, if it affects the germline, is transmissible to subsequent generations. The science of genetics is dependent upon the occurrence and presence of mutations as they allow us to see what happens when a particular system goes wrong, and enable the gene responsible to be identified. Mutation also provides the variation between individuals of the same species and also species to species differences. It follows then that mutations are the raw material of evolution. As a rule, if the mutation has a detrimental effect, it will be removed from the population by the process of natural selection. If the mutation is neutral, i.e. has no phenotypic effect, it may become established in the population by a process of chance. When a neutral mutation reaches a frequency of ⩾1 per cent in the population it becomes known as a polymorphism. On rare occasions mutations are advantageous, and so are selected for. Such mutations will likely become fixed, i.e. reach a frequency of 100 per cent.

The word 'mutation' is greatly misused, and for the sake of consistency that misuse is continued in this book. The mutation is the event of change in the DNA and the mutant region or mutant base is often referred to as the 'mutation' even after transfer to subsequent generations. It would be more correct to refer to the mutated DNA as a 'variant' in subsequent generations. An example of this misuse is the ΔF508 deletion, the most common cause of cystic fibrosis in caucasians, which is usually called the 'ΔF508 mutation'.

In general when we think of mutations, we think of the disruption of genes and an associated disease state or reduction of fitness. In eukaryotes however, coding sequences may only make up as little as 1 per cent of the genome. The majority of mutations therefore occur outside genes and the vast majority of these, have no phenotypic effect. These silent or neutral mutations are generally of little or no interest to biologists. One class of mutation in noncoding DNA which has been of great interest, is that which causes polymorphism of repeat number in tandem repeat arrays. These arrays, the microsatellites and minisatellites, have no known function (at least at the time of writing) but have made a massive impact in the areas of forensics, genome mapping, and population genetics.

A very frequent, but often forgotten, form of mutation is that which results in an abnormality in the number or morphology of the chromosomes. The detection of this type of mutation is well established and relatively simple, and as such will not be covered in detail.

It is mutations which disrupt gene function that are of the widest interest and which form the focus for this book. In many fields of biological science it is important to recognize mutants and characterize the mutations. For the molecular biologist, it is the nature of the mutation and how it affects protein function that is of most interest. For medical scientists, mutation detection allows disease pathogenesis to be understood at the most fundamental level, permitting improved disease management and treatment. Although the focus of this book is the detection and characterization of mutations at the DNA level, it should not be forgotten that mutations in some genes (particularly metabolic genes) can be diagnosed on the basis of simple laboratory tests. For example, disruption of the human phenylalanine hydroxylase gene which causes phenylketonuria, is detected in a standard neonatal test in which a spot of blood, dried onto filter paper, is overlayed onto a bacterial agar plate. Elevated phenylalanine levels are then detected as loss of bacterial growth inhibition.

Until recently, mutation detection at the DNA level was confined to a relatively small number of people, mainly in the field of human genetics, keen on characterizing disease-causing mutations. However, with the rapid advances in the various genome projects and in cDNA sequencing, an ever increasing number of people are needing to get into mutation detection, mainly with the aim of identifying the gene responsible for a particular disease or phenotype, or ascribing a phenotype to a particular gene. With the rapid advances in gene cloning and sequencing, mutation detection has become a major rate-limiting step in determining gene function. It is therefore an active field, with interest from diagnostic laboratories wishing to use the technology and from research laboratories wishing to improve the technology.

At the time of writing there exists no perfect method for finding mutations, but a variety of imperfect methods, each best suited to different applications. The aim of this book is to guide the newcomer into the assortment of methods for finding mutations. The book should help you understand how each method works and which method is most appropriate for what you are trying to do. The vast majority of mutation detection performed today is carried out on human material, and so this book is inevitably slanted towards mutation detection in humans. If your interest however is in a nonhuman organism, you should not be put off by this, as all the same principles apply to any organism.

1.1 History

The history of mutation detection can be divided into two eras: the pre-polymerase chain reaction (PCR) era and the post-PCR era. The pre-PCR era began only in the late 1970s when gene cloning was becoming an established technique. The prerequisite to finding mutations was for the gene of interest to have previously been cloned and the coding region sequenced. A map of restriction endonuclease target sites (restriction map) of the cloned gene greatly facilitated re-cloning of the gene through the construction of genomic libraries using fractionated completely digested DNA into appropriate vector arms. Thus once the gene had been cloned, it could be re-cloned with relative ease (but still a great deal of work) from individuals believed to be mutant for that gene. Once the gene from the patient or mutant organism was cloned, mutation identification required sequencing the entire coding region of the gene. The prior knowledge of the restriction map aided this to some extent, but the sequencing remained an enormous amount of work.

By the mid 1980s methods had begun to appear which enabled the identification of mutations in cloned and uncloned material. The method of denaturing gradient gel electrophoresis (DGGE) could be used effectively for the identification of mutations in cloned DNA. The ribonuclease, RNase A, could be used to identify mutations in uncloned genomic DNA and RNA, by cleaving hybrid wild-type/mutant molecules (heteroduplexes). In a similar manner, chemicals could be used to cleave DNA heteroduplexes in the chemical cleavage of mismatch (CCM) method. All of these methods required a great deal of work and a high degree of skill and experience to perform. Additionally, the RNase and chemical cleavage techniques required the use of large quantities of radioisotopes.

Around 1987 came the revolution which was to radically change the entire field of molecular biology. The polymerase chain reaction was an invention which enabled the artificial replication of a particular region of DNA many times over (Mullis *et al.* 1986). This 'amplification' effectively eliminated the need to re-clone genes for mutation detection. Once part of a gene's sequence was known, primers could be synthesized, allowing amplification of a given fragment with ease from any number of patients or individual organisms.

The DGGE and CCM methods survived the PCR revolution as they became much easier to perform. In the ensuing years many new methods have been developed to find mutations and to detect known mutations. A further impact of PCR on finding mutations, has been to reduce the need for very high sensitivity of molecule detection. This has permitted the conversion of many methods from radioisotopic detection to other detection systems.

1.2 Mutation diagnosis versus mutation scanning

Mutation detection may be divided into two distinct types and different methods are appropriate for each type. The first is the detection of known mutations. For example the cystic fibrosis (CF) gene, *CFTR*, can carry one of several mutations which frequently lead to CF. Most common is the 'ΔF508' deletion of a single phenylalanine codon, which is present in about 70 per cent of caucasian CF alleles. In screening for cystic fibrosis carriers, the mutation detection can be limited to the analysis of only the known common mutations (which together represent around 90 per cent of CF alleles) and still give a very high degree of confidence of correct diagnosis. This type of mutation detection is therefore termed 'mutation diagnosis'. The various diagnostic methods are described in Chapter 4.

The second type of mutation detection is the detection of unknown mutations. This may be in a gene known to cause a particular disease or phenotype, e.g. Duchenne muscular dystrophy, neurofibromatosis, or in a newly cloned gene which is a likely candidate for a particular disease or phenotype. For this type of mutation detection, known as 'mutation scanning', it is usual that each patient or individual organism has a different mutation. Thus the entire sequence of the gene in question must be analysed in each individual. It therefore comes as no surprise that the mutation scanning methods are more complex than the diagnostic methods. The various scanning methods, which each have their own advantages and disadvantages are described in Chapter 5.

1.3 Application of the polymerase chain reaction

Every method bar one discussed in this book is based on the polymerase chain reaction. In most cases the involvement of PCR is to provide a large quantity of the DNA region of interest from the sample available. This not only aids the visualization of the data, but aids mutation detection in systems in which the mutation might be only weakly differentiated from the wild-type. In many of the techniques, the large amount of DNA created by the PCR allows the result to be visualized on gels by UV illumination or by fluorescence emission. In some cases, isotopes are still required, but this does not necessarily indicate a continuing need for the isotope, but either reluctance to modify the protocol if it is only mildly radioactive or general lack of interest in the method, e.g. RNase cleavage (Chapter 5, Section 2).

In the cases of patients, and organisms which cannot be cultured in the laboratory, the quantity of DNA available is usually limited. In order to perform multiple analyses on each sample it is important that each experiment uses as small an amount of DNA as possible.

Some of the mutation finding techniques depend upon the actual process of amplification rather than analysis of the amplification products. Allele-specific PCR (Chapter 4, Section 2) relies on one PCR primer lying over the mutation site and so amplification occurs or not, depending on the presence or absence of the mutation. The 5′ nuclease assay (Chapter 4, Section 6) is a system in which a fluorescence signal reflects the extent of DNA synthesis across a mutant site during the PCR process.

All of these techniques therefore require that you can reliably perform PCR reactions which give good yields of specific product without co-amplification of nonspecific background products or products resulting from contamination.

1.4 Outcomes of mutation

Mutations affecting gene function may be divided into two types: regulatory and structural mutations. Regulatory mutations lie outside of the gene's coding region and affect the expression. Structural mutations are mutations which alter the amino acid of the protein product. The general focus on mutations and the focus of this book, is on structural mutations. Structural mutations in which a single base is replaced by another (point mutations), can have three possible outcomes:

(1) If the change causes the relevant codon to still encode the normal amino acid (a synonymous change), the mutation is silent;
(2) if the mutation causes the codon to encode a different amino acid (a nonsynonymous change), the mutation is known as a 'missense' mutation; and
(3) if the mutation alters the codon to a termination codon (i.e. TAA, TAG, or TGA), the mutation is known as a 'nonsense' mutation, and the protein produced is prematurely truncated at the point corresponding to the position of the mutation.

The most common point mutation is the substitution of a cytosine (C) by a thymine (T) in a CpG nucleotide pair. This is because methylated Cs in this nucleotide pair are readily deaminated to form T. This relative instability results in a general deficiency of CpG nucleotide pairs outside of unmethylated so-called CpG islands (Cross and Bird 1995).

Mutations which insert or delete multiples of 3 bp maintain the reading frame and so result in the lengthening or shortening of the protein without altering the sequence except at the insertion/deletion site. If the insertion or deletion is not a multiple of 3 bp, a shifting of the reading frame will occur (a frameshift), and the protein produced will be grossly abnormal from the point of the mutation to the carboxyl terminus and be shortened or lengthened, depending on where a new termination codon happens to be situated.

The mutation detection techniques described in this book will provide a basis for choosing an appropriate method for detecting structural mutations. A form of mutation affecting gene function that would usually be missed by these techniques, is the regulatory mutation causing loss of gene expression, known as the null mutation. Null mutations are usually mutations in the gene's promoter, or occassionally in intronic enhancer sequences. It is unclear how common phenotype-producing null mutations are, but they are probably very rare. Therefore, as a general rule in mutation detection, only when a screen for mutations in a gene which is certain to produce a given phenotype fails to find any abnormality, should you begin to consider the possibility of null mutations.

2. Mutation nomenclature

Just as there has been no single method for the detection of mutations, there has been no single system for the naming of mutations once they have been characterized.

It is possible to divide mutation nomenclature into two parts, based on those mutations which are visible under the microscope (i.e. chromosomal rearrangements) and those which are submicroscopic.

2.1 Chromosomal rearrangement nomenclature

2.1.1 Translocations

The chromosomal rearrangement mutations have a well established nomenclature system. For example the human rearrangement: 46,XY,t(1;11)(p36.3;q13.1) indicates a male individual with a balanced translocation between chromosomes 1 and 11 with a breakpoint on the short arm of chromosome 1 at position 36.3 and a breakpoint on the long arm of chromosome 11 at position 13.1. The positions refer to the chromosomal banding pattern. In this case, the translocation is balanced, i.e. the person has one normal copy of both chromosomes 1 and 11 plus the translocated chromosomes. Thus overall, he has two copies of all parts of chromosomes 1 and 11. If this individual were to father a child and pass on one of the rearranged chromosomes, the child would be 'unbalanced' for the translocation as it would not have two copies of all parts of chromosomes 1 and 11. In other words, the child would have one copy of part of one of the chromosomes concerned and three copies of part of the other chromosome concerned. The karyotypic description of the child would depend on which affected chromosome he or she inherited from the father, but could for instance be 46,XX,der(11)t(1,11)(p36.3;q13.1), meaning that the bulk of the abnormal chromosome is derived from chromosome 11. In this case the child would have three copies of the tip of the short arm of chromosome 1 and one copy of the tip of the long arm of chromosome 11.

2.1.2 Inversions

Inversions can occur on a chromosome when two breaks occur and the piece between the breakpoints is reversed in orientation. In the example: 46,XX,i(17)(p22;p11.1), the inversion occurred on chromosome 17 of a female individual, with breakpoints at positions 22 and 11.1 on the short arm.

2.1.3 Deletions

Loss of fragments of a chromosome can occur either from within the middle of a chromosome arm (an interstitial deletion) or from the end of the chromosome (a terminal deletion). The description: 46,XY,del(9)(p23.1;p24.1) indicates an interstitial deletion from the short arm of chromosome 9, whereas the description: 46,XY,del(9) (p23.1;pter) indicates a terminal deletion on the short arm of chromosome 9.

2.2 Submicroscopic mutation nomenclature

Submicroscopic mutations fall into three categories: point mutations, deletions and insertions.

2.2.1 Point mutations

Point mutations are the substitution of a single base pair, e.g. a G to A point mutation changes a G–C base pair to an A–T base pair. Point mutations can be subdivided into transitions and transversions depending on what base is replaced with what. A transition is the replacement of a purine with a purine, whereas a transversion is the replacement of a purine with a pyrimidine or vice versa.

Examples of transitions:
A (purine)   →   G (purine)   i.e. A–T → G–C
T (pyrimidine)   →   C (pyrimidine)   i.e. T–A → C–G

Examples of transversions:
A (purine)   →   C (pyrimidine)   i.e. A–T → C–G
T (pyrimidine)   →   G (purine)   i.e. T–A → G–C

Much of the early work on human mutation detection and characterization was centred on the thalassaemia blood disorders and the causative haemoglobin genes. Prior to the days of mutation characterization, detectably abnormal proteins acquired names such as haemoglobin S (Hb^s), the abnormal haemoglobin causing sickle-cell anaemia. It was also common for the abnormality to be named after the location where the first clinical case was described, e.g. Hb Constant Spring, Hb Kenya.

When the underlying mutations began to be characterized, it was common for them to be described as for example, nonsense mutation in codon 121 (GAA → TAA), but not for them to be given an abbreviated name. Such a system requires that you know the position of the initiator methionine, which represents amino acid or codon position 1.

Fig 1.1

Gene and protein numbering system. The first base of the gene to be transcribed is termed '+1', the second is '+2' etc. The base pairs of the promoter have a minus number, in which '–1' is the base pair immediately 5' to '+1'. The N-terminal amino acid of the protein is amino acid number 1.

A convention is now emerging in which the amino acid/codon number and protein change is indicated. In the initial publication, the mutation might be described as 'an isoleucine to methionine substitution at codon 135 (ATC → ATG)'. This might then be abbreviated using the three-letter amino acid abbreviation system to Ile-135 → Met, and referred to as this in any subsequent discussion or publications.

More commonly the one-letter amino acid abbreviation system is used, so that the same mutation would be described as I135M. Occasionally this is also written as I135M (C → G) to indicate the DNA change. In this nomenclature system, nonsense mutations are indicated as 'X', so that a termination codon at position 135 would be represented as I135X.

2.2.2 Deletions and insertions

Deletions and insertions are the loss or gain of anything upwards of a single base pair, respectively. Deletions are indicated as 'Δ' (or 'delta'), so that an in-frame deletion, for example causing the loss of amino acid 135, would be represented as ΔI135. Deletions or insertions causing frameshifts, are indicated noting the position of the change in the DNA rather than the protein, so that for example a 2-bp GT insertion at codon 135 would be represented as 438insGT, where 438 indicates the base number, relating to base 1—the first transcribed base, also known as the cap site. Similarly a 2-bp GT deletion at the same position would be represented as 438delGT. When the insertion or deletion is longer than about 3 bp, it is usual to indicate the size of the change rather than the sequence, e.g. a 14-bp insertion at DNA position 438, would be represented as 438ins14. Complex mutation events such as a substitution or insertion accompanying a deletion, for example the deletion of ATA at DNA position 4114–4116 and replacement with TT, is represented as 4114ATA → TT.

2.2.3 Splice mutations

Splice mutations are another class of mutation, affecting multi-exon genes. Mutation of the sequences (by point mutation, insertion, or deletion) at the 5′ (donor site) and the 3′ (acceptor site) of the intron can prevent correct splicing of the pre-mRNA, leading to a grossly abnormal protein. Alternatively, mutations within exons can create cryptic splice sites, with much the same result. Consistently, the donor site sequence begins with 5′-GT-3′ and the acceptor site sequence ends with 5′-AG-3′. Approximately 15 per cent of point mutations which cause human genetic disease, do so by affecting pre-mRNA splicing.

A mutation disrupting the donor site, results in the intron not being excised from the RNA molecule. Thus protein translation occurs into the intron, incorporating a new amino acid sequence and terminating at the first stop codon. A mutation disrupting the acceptor site causes 'exon-skipping' as the nearest available acceptor site (at the next intron) is used instead.

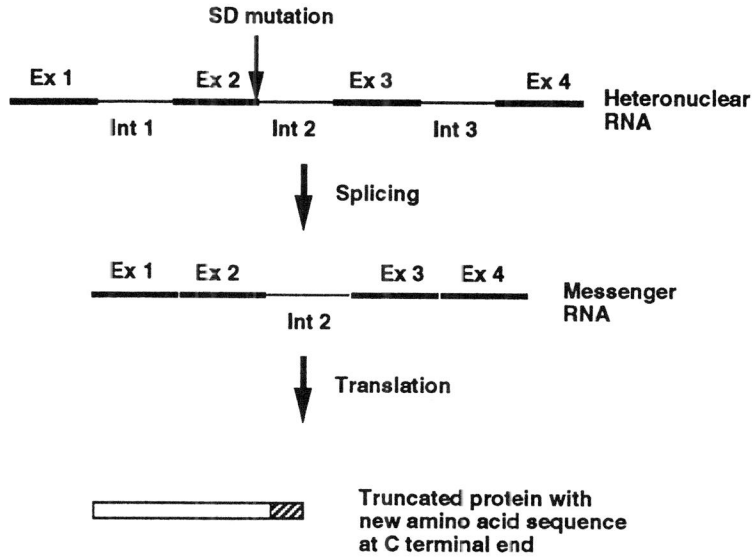

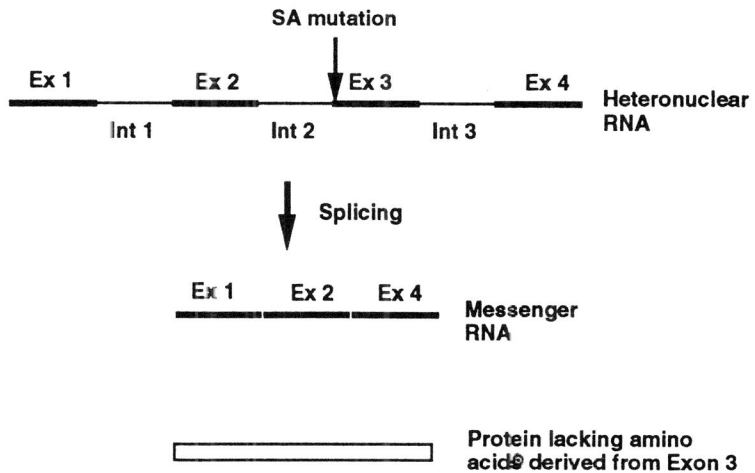

Fig 1.2

The effect of mutations in splice donor (SD) and splice acceptor (SA) sites. Splice donor mutations cause the retention of the intron, whereas splice acceptor mutations cause the deletion of an exon.

3. Experimental considerations

If there was one perfect method for mutation diagnosis or mutation scanning, there would be no need for this book. But unfortunately, there are no ideal methods. Some methods are very difficult to perform and some are relatively poor at detecting mutations, i.e. will fail to detect some mutations. Amongst all the techniques, there is a tendency for the more difficult techniques to have the highest mutation detection efficacies and the easiest to have the lowest efficacies.

3.1 Efficacy of detection required

Before you start, you need to decide exactly what it is you want to achieve. You might be using the technology to detect polymorphism, to prove a candidate gene to be a disease gene, to get an idea of how the mutations are distributed through the gene, or, if in a clinical laboratory, to detect the disease-causing mutation in every patient sample.

If you are looking for polymorphism it is likely that you can afford to miss a fair number of possible variants. If you are trying to prove a candidate gene to be a disease gene, you only need to detect one mutation and can therefore get away with a relatively low detection rate. However, if you only have a few patient samples, it then becomes important to have a high detection rate.

It is clear then, that you might be happy to use a technique that misses some mutations, if it is easy and cheap to perform. Alternatively, if you need to detect all mutations you may have to opt for a method which is relatively difficult to perform.

3.2 Availability of time, personnel, and specialist equipment

Apart from the choice to be made about the detection efficacy required, choices have to be made regarding the cost, the time available, the number of personnel available, and their level of experience and expertize. You might also be limited in what equipment you have at your disposal. For example, if your laboratory is set up for DNA sequencing, you should be able to do techniques such as the single-strand conformation polymorphism assay or chemical cleavage of mismatch, but not denaturing gradient gel electrophoresis which requires specialist equipment (see Chapter 5, Sections 1, 3, and 4). Some laboratories are able to buy any equipment which they would like, but yours is probably not one of these. You might therefore be limited by your laboratory environment as to what techniques and level of efficacy are available to you.

3.3 Safety considerations

Another level of choice limitation is dictated by safety considerations. The main considerations are to do with the handling of radioisotopes and toxic chemicals. You must first decide whether your radioisotope manipulation areas and fume hood facilities are suitable for the technique(s) you have in mind. If not, you must choose either to improve the facilities or to use a different technique which avoids the problem. You must also decide whether you, or whoever it is doing the experiments, have the training required to handle such substances safely.

3.4 Overall strategy

Which approach you take largely depends on whether you have no clue as to the location of the gene responsible; have a map position, but have not characterized the gene; or have characterized the gene, or expect a particular type of mutation, e.g. nonsense mutations in certain oncogenes (see Chapter 5, Section 6) or triplet repeat expansion (see Chapter 3, Section 3.3).

If you have no idea of the map position of the gene, you are probably going to be looking for large chromosomal rearrangements. To do this, you will need the relevant microscopy equipment and to have living cells available for karyotyping (see Chapter 3, Section 1).

If you have a map location, based on say a translocation breakpoint or deletion, you might be looking for more subtle rearrangements to further pin-point the gene. In this case, you will probably need to perform Southern blots of ordinary agarose gels or pulsed-field gels. If you have some sequence information of the gene and have a probe, small re-arrangements may also be revealed by Northern blotting if the relevant tissue is available from the subject in question. (for details, see Chapter 3, Section 2).

If the mutant gene has been characterized, direct analysis of the mutation will be required. The type of analysis depends on whether you need a mutation diagnosis method or a mutation scanning method. The method chosen will depend on your laboratory circumstances and the detection efficacy required (see above). Details of the various diagnostic methods and scanning methods along with their specific advantages and disadvantages are given in Chapters 4 and 5, respectively. Newer methods for mutation detection which have not yet been thoroughly evaluated are described in Chapter 6.

References

Cross, S. H. and Bird, A. P. (1995) CpG islands and genes. *Current Opinion in Genetics and Development*, **5**, 309.

Mullis, K. B., Faloona, F. A., Scharf S. J., Saiki, R. K., Horn, G. T., and Ehrich, H. A. (1986). Specific enzymatic amplification of DNA *in vitro*: the polymerase chain reaction. *Cold Spring Harbor Symposium on Quantitative Biology*, **51**, 263.

2 Logistics

1. Substrate for analysis: DNA or RNA?

To the molecular biologist, DNA comes in two forms, genomic DNA and cDNA. If your organism of interest is eukaryotic, genomic DNA consists of the nuclear chromosomal DNA and the cytoplasmic mitochondrial DNA. Genes can be very large, depending on the number and size of introns. In humans an 'average' gene has a coding sequence of about 1500 bp, but this will usually be distributed throughout several exons, spanning several kilobases of DNA. However, some genes may consist of just a single exon and others may consist of very many small exons spread over many scores or even hundreds of kilobases of DNA.

cDNA or 'complementary DNA' is the DNA copy of messenger RNA transcripts contained within the cell. In order to clone or PCR-amplify RNA transcripts, it is first necessary to convert the RNA back into DNA. This is done using reverse transcriptase, an enzyme used by retroviruses to convert their RNA genomes into DNA. As messenger RNA (mRNA) has had the intronic sequences removed by the process of splicing, cDNA copies of genes are significantly smaller than the genomic equivalents and are therefore easier to analyse.

It would therefore appear for the purposes of mutation detection, that cDNA is a more desirable substrate than genomic DNA. Sadly, the situation is more complicated than this with some genes being best approached via cDNA and others best approached via genomic DNA. The reasons for this are mainly due to the structure of the gene in question and the availability of its transcripts.

1.1 Availability of transcripts

The availability of transcripts is dependent on the expression profile of the gene in question. Genes involved in respiration and the maintenance of cell structures and materials, the so-called house-keeping genes, are expressed equally in all tissues. Most genes however, are expressed in only some tissues and in different tissues at different times of development. If the organism you are working on is an animal which should not be harmed by the mutation screening investigation, there is a fair chance that the gene of interest will not be expressed in an accessible tissue, e.g. skin or blood. If the organism is expendable, e.g. *Drosophila*, the expression profile is not such a concern.

If you have access to a particular tissue and the gene of interest is highly or moderately expressed in that tissue, cDNA offers very real advantages over genomic DNA, especially if the gene is large and divided into many small exons. If the gene is expressed in that tissue at a low level, well designed PCR primers may still allow reliable amplification. Many genes, however, are not expressed in accessible tissues and are therefore a problem if the gene is large and divided into many small exons.

1.2 Leaky transcription

It has become apparent that in tissues where a given gene is 'not expressed' a very low level of transcription does in fact occur, providing a potential source of cDNA (Sarkar and Sommer 1989). This phenomenon is usually known as 'illegitimate' or 'ectopic' transcription. Both descriptions indicate that this very low level transcription is a biological mistake or error. It is more probable though, that the mechanisms for controlling gene expression do not allow complete silencing of gene expression. Therefore, the terms 'leaky transcription' or 'trace level expression' are probably more appropriate. Leaky transcription probably produces in the region of one transcript per 1000 cells, making the RT-PCR amplification a nontrivial exercise.

Such amplifications (usually performed on lymphocyte cDNA) will always require two rounds of PCR amplification. Following the primary round of amplification (30–35 cycles), 1 μl of product should be diluted at least 50-fold and then subjected to a secondary round of amplification (again 30–35 cycles), but this time using a second pair of primers, nested within the first primer pair. This should yield enough product for any mutation detection experiment. Extra care however should be taken to avoid cross-contamination when conducting two rounds of amplification.

1.3 Substrate choice in mutation diagnosis

In cases where you are diagnosing individuals for a single known mutation, the choice of substrate will always be genomic DNA as you will only need to amplify the relevant exon or region. In cases however, in which you are characterizing one of several possible mutations in a gene, e.g. the cystic fibrosis gene, you may need to examine several exons. In such a case, cDNA might be the better substrate as it may be possible to perform only a single amplification per individual, rather than needing to perform a separate amplification for each exon of interest.

1.4 Potential pitfalls associated with cDNA

As a point of caution, it should be noted that RT-PCR products derived via leaky transcription may not have the desired sequence structure if the transcript is normally the subject of tissue-specific splicing, i.e. the alternative splice form of interest may be absent from the RNA source available. It has also been observed that 'extra' exons can be included in leaky transcripts which are not normally included in any tissue-specific transcript (Roberts *et al.* 1993). It is therefore important to exercise caution when conducting such work and not to infer the gene structure from information derived by this route.

If cDNA has been used as the substrate, and regardless of whether there was an abundant or leaky level of transcription in the tissue or cell type used, nonsense mutations can be a cause of false mutation diagnosis. Some nonsense mutations can cause a great reduction in the levels of mRNA derived from the mutant allele (McIntosh *et al.* 1993). In cases where the mutation is heterozygous, this can easily result in the mutant allele being missed and thus giving the impression of being normal. An additional rare occurrence is the splicing out of exons containing nonsense mutations (Dietz *et al.* 1993). In such cases an exon which is normally present in all alternative splice forms, is absent from the mRNA despite all the sequences involved in splicing being unaltered. The mechanism of how nonsense mutations alter splice site selection is unclear.

Together, these potential pitfalls go to stress the need for caution when using cDNA as a substrate in mutation detection. If it is essential that you have a 100 per cent correct mutation diagnosis, then it is probably advisable not to use cDNA as the sole substrate.

1.5 Significance of gene structure

As outlined above, when your gene of interest is large, with many perhaps small exons, cDNA has many advantages over genomic DNA as a substrate. A further advantage is that splice site mutations should produce a PCR product of an abnormal size, eliminating the need to screen for such mutations. If, however, you have chosen or are

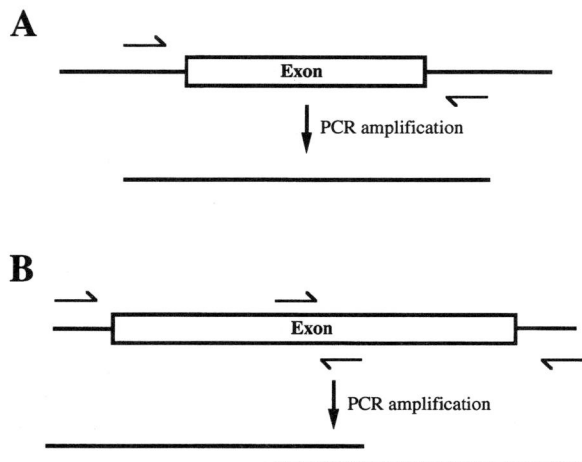

Fig 2.1

Exon amplification from genomic DNA. A, an exon which can be amplified as a single product. B, an exon which is too large to be amplified as a single fragment. In this case, the PCR products should be overlapping, so a mutation lying under one primer will not be missed.

restricted to genomic DNA, it is important to situate PCR primers for whole exon amplifications in the flanking intronic DNA. Situating the primers at the ends of the exon would miss the splice sites at the intron/exon junctions and also miss any mutation lying under the primers themselves (see Chapter 5, Section 1.1.2). If possible, the primers should be placed so that about 30 bp of intronic DNA is amplified on each end of the PCR fragment.

In addition to screening the entire coding region and splice sites of the gene, it may be desirable for the sake of completeness to test the promoter region, the 3′ untranslated region (3′ UTR) and (if eukaryotic) the polyadenylation site or sites. Although genes may have control elements lying many kilobases 5′ and/or 3′ to the gene, the main promoter elements will usually lie within a few hundred base pairs of the transcription start site. Many such mutations have been identified in the mammalian globin and factor VIII genes. The 3′ UTR in some cases appears to be involved in mRNA stability or transport and is therefore worth including in any screen providing it does not require much extra work. In rare examples, the 5′ UTR can contain regulatory sequences and is therefore also worth including. The polyadenylation signal is the sequence (usually AATAAA) towards the 3′ end of the mRNA which instructs the eukaryotic cell to add a tail of adenines to the 3′ end of the molecule. This has the effect of stabilizing the molecule. Mutations abolishing the polyadenylation signal can cause RNA instability and therefore a reduction of the level of protein product.

2.   Analysis of old material

Mutation detection need not be limited to specifically obtained DNA samples. The ubiquitous use of amplification methods in mutation detection allows DNA to be obtained from hospital-preserved tissue samples, from microscope slides, and even from ancient archaeological specimens.

In western countries a high proportion of individuals at some time in their lives have tissue samples taken, analysed, and stored. This permits the retrospective analysis of DNA sequences of either the patient's DNA or the DNA (or RNA) of infectious agents contained within the sample.

2.1   Paraffin embedded samples

The most common method for preserving and storing human tissues is fixation in formalin followed by embedding in paraffin. DNA may be successfully extracted from samples prepared in this way several decades previously, and also from sections made from the embedded sample. The DNA extracted from such tissues is always degraded to some extent. In a typical case, when the extracted DNA is run out on an agarose gel, a smear is seen in which the majority of the DNA falls in the 100 to 800 bp size range. It is therefore desirable to amplify only small fragments. In general, there is an inverse relationship between the efficiency of amplification and the size of the amplification fragment.

The process of DNA extraction from paraffin blocks usually involves the sectioning of the blocks, the trimming away of as much paraffin as possible followed by extraction in octane or xylene to remove the remaining paraffin. When DNA is to be extracted from stained sections on microscope slides, the slides should be soaked in xylene to facilitate removal of the coverslip and the tissue scraped off with a razor blade. At all stages utmost care must be taken to avoid contamination from either the worker or other samples. The tissues are next treated with proteinase K, and then heated at 95 °C to inactivate the proteinase K. The crude DNA obtained is suitable for PCR amplification. Differing quantities of sample DNA should be subjected to amplification and a high number of cycles, each with relatively long times at each temperature, should be performed.

Successful amplification of such material is not guaranteed. Samples which amplified immediately after DNA extraction may fail to amplify after a period of frozen storage. Also some samples may never amplify. It is possible that this phenomenon is due to the particular method and time period of fixation.

3. Preparation and storage of samples and reagents

The primary concerns in the obtaining and handling of samples for analysis, whether they be blood, tissue, or the whole organism, are that the samples should be isolated to avoid contamination and maintained in an environment which will not cause degradation of the sample.

In most cases, for mutation detection, DNA samples are derived from separated white blood cells, whole blood, cultured cells, or tissue biopsies. RNA may also be obtained from these samples, but when blood is used, white cells should be separated. White cells are readily separated by density gradient centrifugation, using commercially available reagents such as 'Ficoll-Paque' (Pharmacia Biotech.).

When viral DNA or RNA is to be extracted from blood or cultured cells, to enrich for viral particles the material should first be centrifuged at low speed, the supernatant then centrifuged at a moderately high speed to remove remaining large particulates, and then the supernatant centrifuged at very high speed to pellet viral particles.

3.1 Storage of tissue samples

In all cases, DNA should be extracted when the material is as fresh as possible. When RNA is extracted it is of utmost importance that the material is ultra fresh, has been stored for a very short period on ice, or has been snap-frozen in liquid nitrogen. In certain situations it is not possible to process the material immediately or to chill or freeze it, e.g. when patient blood samples are sent to appropriate laboratories for analysis or when material is obtained during field work. Solid tissue may be stored reasonably well in 50 per cent ethanol or in a saturated salt solution containing 20 per cent DMSO without the need for refrigeration. Blood may be stored long term by simply air drying smears on microscope slides or by drying onto a specialized support such as 'IsoCode Stix' (Schleicher and Schuell).

3.2 Storage of DNA and RNA samples

◇ 1 × TE is 10 mM Tris–HCl (pH 7.6), 1 mM EDTA.

Genomic DNA pellets may take several days to fully dissolve in water or TE, and so should be maintained at room temperature or 4 °C for some time before freezing. It should be noted that repeated freeze/thaw cycles may damage the DNA. If the DNA has been extracted with care, it should be stable at 4 °C for several years.

Due to its lower molecular weight, RNA should dissolve within minutes. The RNA should be stored long term at –20 °C or more safely at –70 °C. As RNA degrades readily, it is best when removing an aliquot of RNA from the stock, to allow the stock to thaw only enough to allow the removal of the desired amount, and then to return it to the freezer.

3.3 Storage of PCR reagents

All PCR reagents, i.e. buffer, nucleotides, primers, should be stored aliquoted into several small samples. If contamination or PCR failure occurs, the 'active' aliquots should be discarded and new ones started. The buffer should be kept frozen, but should not suffer from repeated freeze/thaw cycles. The nucleotides should be frozen, but may suffer from repeated freeze/thaw cycles. This component should therefore be kept in very small aliquots. The primers may also suffer from freeze/thaw cycles, but should be stable at 4 °C for long periods, especially if the pH is maintained around neutral or if they are maintained in 1 × PCR buffer. It is therefore advisable to store aliquots of primer at –20 °C, but to keep the 'active' aliquot at 4 °C.

Further reading

Wright, D. K. and Manos, M. M. (1990). Sample preparation from paraffin-embedded tissues. In *PCR protocols: A guide to methods and applications.* (ed. M. A. Innis D. H. Gelfand, J. J. Srinsk, and T. J. White), pp. 153–8, Academic Press, San Diego, CA. A good protocol for amplification from paraffin-embedded tissues.

References

Dietz, H. C., Valle, D., Francomano, C. A., Kendzior, R. J., Pyeritz, R. E., and Cutting G. R. (1993). The skipping of constitutive exons *in vivo* induced by nonsense mutations. *Science*, **259**, 680.

McIntosh, I., Hamosh, A., and Dietz, H. C. (1993). Nonsense mutations and diminished mRNA levels. *Nature Genetics*, **4**, 219.

Roberts, R. G., Bentley, D. R., and Bobrow, M. (1993). Infidelity in the structure of ectopic transcripts: a novel exon in lymphocyte dystrophin transcripts. *Human Mutation*, **2**, 293.

Sarkar, G. and Sommer, S. S. (1989). Access to a messenger RNA sequence or its protein product is not limited by tissue or species specificity. *Science*, **244**, 331.

3

The detection of rearrangements

The bulk of this book describes approaches for the characterization of point mutations and very small rearrangements. It must however not be forgotten that a mutation can be a gross rearrangement, sometimes involving the gain or loss of whole chromosomes. Chromosomal aberrations are in fact very common, accounting for 0.6 per cent of human live-births and up to 5 per cent of still births. The single most common mutation amongst humans being the nondisjunction of chromosome 21 in meiosis, causing Down's syndrome.

The approaches to the detection and characterization of re-arrangements vary according to the size of DNA involved. Very large rearrangements can be visually detected by karyotyping using conventional chromosome banding methods and finer resolution can be obtained by fluorescent *in situ* hybridization. Smaller rearrangements can be detected and analysed by the Southern blotting of either pulsed-field or conventionally run agarose gels. Intra-gene rearrangements may be detected by Northern blotting, by RT-PCR amplification or by multiplex PCR amplification of individual exons. Dynamic mutations involving the expansion of triplet repeats are detected as abnormally sized alleles on gels following PCR amplification.

Which of these methods you should use will depend on whether you are investigating a candidate chromosomal region or gene, or have an indication of the type of mutation to expect.

1. The detection of large chromosomal rearrangements by microscopy

A large number of human genetic diseases are caused by chromosomal abnormalities which are visible under the microscope, i.e. aneuploidy (an abnormal number of chromosomes) or rearrangements such as deletions, inversions, and translocations. There are several microscopy-based methods which can be used for the detection of rearrangements and each has a different level of resolution.

1.1 Chromosome banding

In the early 1970s developments were made which allowed mammalian metaphase chromosomes to be stained to produce a banding pattern specific for each chromosome. This banding pattern enabled cytogeneticists to differentiate between chromosomes of similar size. Not only was it possible to determine which chromosome was trisomic or monosomic in cases of aneuploidy, but also to recognize translocations and determine which chromosomes were involved. The most common methods to identify mammalian chromosomes are G (Giemsa), R (Reverse), and Q (Quinacrine) banding. All depend upon the same principle of chromatin denaturation and/or enzymatic digestion followed by incorporation of a DNA-specific dye. The pattern of banding produced, is a reflection of metaphase chromosome structure in which the DNA is packaged >10 000-fold. It is also becoming clear that there is a correlation between gene density, base composition, repeat sequence content, replication timing, and the banding pattern obtained.

Despite the ability to perform high-resolution banding (850 bands on the human metaphase genome) the visibility of chromosomal rearrangements is limited to events involving regions of DNA greater than about 5 Mb (5 000 000 bp). However, recent developments in the areas of *in-situ* hybridization and the visualization of nonmetaphase chromosomes have greatly increased the resolution of visualization of gross structural mutations.

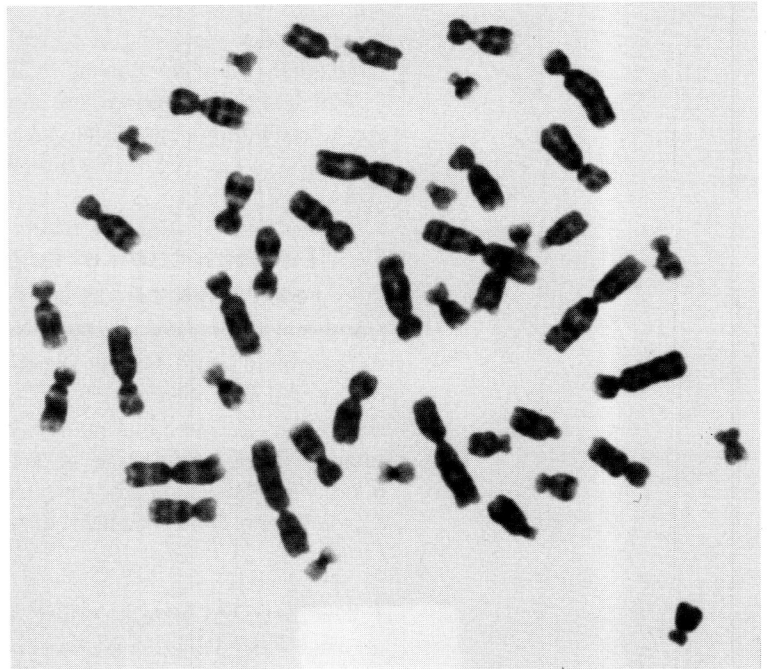

Fig 3.1

G-banded human metaphase chromosomes.

1.2 Fluorescent in-situ hybridization

The technique of fluorescent *in-situ* hybridization (FISH) has made a massive impact on gene mapping, but it is also useful in the high-resolution analysis of chromosomal rearrangements. The high number of markers now available enable the end points of relatively small rearrangements to be finely localized. Probes can consist of any plasmid, bacteriophage, or cosmid clone and are labelled with haptens such as biotin or digoxigenin. Following hybridization to the chromosomes, the probe is visualized by fluorescence. When performed on metaphase chromosomes, this technique allows rearrangements to be visualized with a resolution limit of about 1 Mb.

An additional benefit of FISH has been the production of 'chromosome paints'. The 'paints' are derived from repetitive sequences derived from a particular chromosome, by repeat sequence PCR amplification from chromosome-specific libraries or flow-sorted chromosomes. The combination of five different fluorochromes, complex probe labelling to produce a high number of paint 'colours', as well as the development of microscope filters and computer software, have enabled the production and co-visualization of paints specific for each of the human chromosomes. This technology is applicable to most types of rearrangement, except those involving only a single chromosome, e.g. small deletions or paracentric inversions.

The similar technique known as 'reverse chromosome painting' involves the flow-sorting of the abnormal chromosome from the affected individual. The abnormal chromosome is then used to derive a 'paint' which is hybridized back to normal chromosomes. In this way deletions, insertions, and translocations can be readily diagnosed.

1.3 In-situ hybridization to nonmetaphase chromosomes

The limit of resolution obtained with FISH can be increased by using the significantly less condensed interphase chromosomes. The use of interphase chromosomes is a relatively new phenomenon as conventional banding is limited to metaphase chromosomes. FISH with interphase chromosomes allows differently coloured probes lying as close as 50 kb to be resolved. Thus this method can permit the very fine definition of chromosomal rearrangement end-points and can even arrange overlapping cosmid clones spanning a breakpoint into a contig.

Fluorescent *in-situ* hybridization to free chromatin fibres released from lysed cells provides an alternative form of analysis, allowing probes lying as close as 20 kb to be resolved. A further increase in resolution to as little as 5 kb can be obtained by preparing streams of linearly extended DNA across a microscope slide. This technique though, has as yet no advantages over Southern blotting for the characterization of small rearrangements.

2. The detection of small rearrangements by blotting

2.1 Southern blotting

Southern blotting is the transfer of size-separated restriction endonuclease digested DNA from (normally) an agarose gel to a membrane. Specific DNA fragments are identified when the membrane is hybridized with a specific probe. Conventionally run agarose gels will resolve DNA fragments within the 500 bp–20 kb size range. This type of analysis will therefore permit the characterization of rearrangements falling within this size range or the characterization of the junctions of larger rearrangements. Rearrangements falling in the 20 kb–2 Mb size range therefore require an alternative method.

Such a method was devized in the 1980s and is known as pulsed-field gel electrophoresis (PFGE) (Schwartz and Cantor 1984). In conventional electrophoresis, DNA molecules pass through the pores in the gel matrix, the smaller molecules passing through the pores more easily than the larger molecules, resulting in size separation of the DNA fragments. The gel has a resolving limit because the fragments that are larger than the average pore size migrate in a manner independent of their size, travelling in a snake-like motion. In the most common form of PFGE called contour-clamped homogeneous electric field (CHEF), the electrodes are spaced in a hexagonal array around the gel (Vollrath and Davis 1987). The electric field is switched back and forth (pulsed) from a 'northwesterly' direction to a 'northeasterly' so that the DNA fragments zigzag down the gel. Each time the electric field switches, the molecules have to reorientate themselves into the new direction before they begin to migrate again. The larger the molecule, the longer the time taken to reorientate is, and so very large molecules are enabled to migrate according to their size.

PFGE therefore enables you to cut DNA with rare-cutting restriction enzymes, so that you can analyse relatively large regions of DNA for rearrangements.

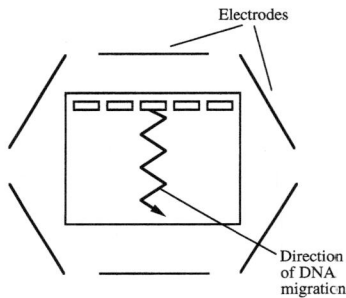

Fig 3.2

DNA migration through a CHEF gel.

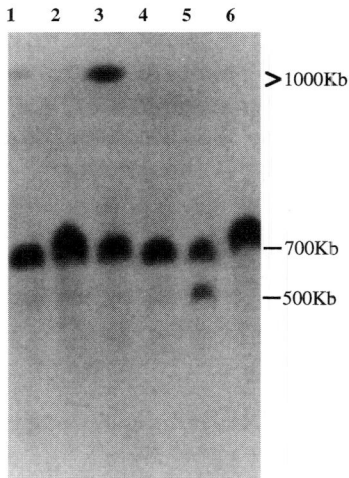

Fig 3.3

A translocation breakpoint detected as an abnormal sized band on a CHEF–pulsed-field gel Southern blot. (Reproduced with permission from Zelenetz et al. 1991; W. B. Saunders Co.)

Table 3.1 Resolution of rearrangement detection and analysis techniques

Technique	Resolution
Banded metaphase chromosomes	>5 Mb
FISH on metaphase chromosomes	>1 Mb
FISH on interphase chromosomes	50 kb–1 Mb
FISH on extended DNA	5 kb–700 kb
Pulsed-field Southern blots	20 kb–2 Mb
Ordinary Southern blots	500 bp–20 kb
Northern blots	500 bp–15 kb
PCR	<2 kb

2.2 Northern blotting

Northern blotting is the transfer of size-separated RNA from a denaturing agarose gel to a membrane. Specific transcripts are identified when the membrane is hybridized with a specific (usually DNA) probe. In some cases of mutation detection, there is access to tissue in which the mutant gene is expressed. If a rearrangement within the transcribed region of the gene is present and significantly alters the length of the transcript, it should be detectable by Northern blotting. Although only a very small proportion of mutations may be identified by either Southern or Northern blotting, this route should not be forgotten as it provides a rapid and simple preliminary screen.

3. Detection of small deletions, insertions, and expansions

Some genes, e.g. the gene for Duchenne muscular dystrophy (DMD) in humans, show a preponderance of deletions, rather than point mutations. In such cases it is therefore desirable to perform a preliminary screen for size alterations within the gene before embarking on point mutation detection. As only relatively large intra-genic rearrangements may be detected by Southern and Northern blotting, PCR is a more attractive approach as it can be used to detect all but the smallest of deletions and insertions.

3.1 RT-PCR amplification

In some cases it will be possible to obtain tissue in which the gene under investigation is expressed and to prepare RNA (see Chapter 2, Section 1.1). When this is possible, the entire transcript may be amplified as a single product or as several overlapping products. These may be sized on agarose or acrylamide gels and determined to be of the normal size or not. This approach has the advantage over genomic amplification (see below) that exon deletions or intron retentions caused by splice site point mutations will also be detected.

3.2 PCR amplification of exons

In many cases access to transcripts is not possible or is too difficult. In these cases it may be desirable to amplify individual exons or parts of exons from genomic DNA. This approach has been very successful in the study of mutations in DMD. Several exons may be PCR amplified simultaneously in the same tube (multiplex amplification). Each amplification product is designed to be of a different size, so that when the products are run on an agarose gel, a ladder of bands is seen.

Fig 3.4

Exon deletions in the human *HPRT* gene, detected by multiplex PCR. Each band represents one exon, except for the largest band which represents two exons. Lanes 5 and 6 contain the amplification products from normal individuals. The samples in lanes 1, 2, 3 and 4 are derived from individuals with Lesch–Nyhan syndrome. The sample in lane 2 has a complete gene deletion whereas the samples in lanes 1, 3, and 4 have deletions of one or more exons. (Reproduced with permission from Gibbs *et al.* 1990; Academic Press, Inc.)

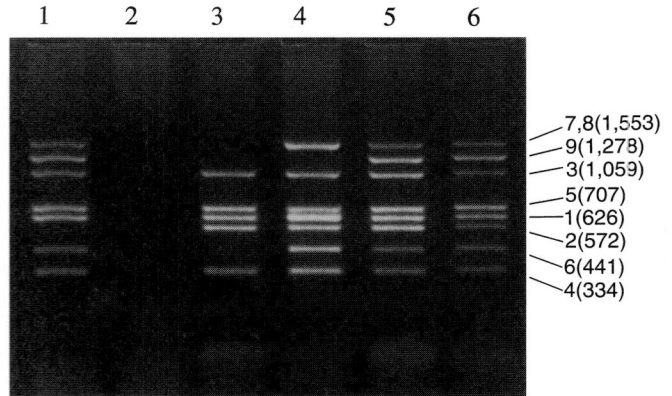

Exon deletions are simply detected by the absence of a particular band on the gel.

The multiplex PCR approach is unfortunately not as simple as it may seem. If seven fragments of DNA are to be co-amplified, all 14 primers must work well and not interfere with any of the other primers. So not only must each primer pair have approximately the same melting temperature, but the amplification conditions must be optimized for all primers together. This may involve some primers needing to be used at a different concentration to the others.

Unlike DMD which lies on the X-chromosome and is single-copy in boys, most genes in diploid organisms have two alleles. In these cases, visualization of DNA fragments on ethidium bromide-stained agarose gels does not usually allow the determination of target DNA dosage. Efforts to overcome this problem have included: using a low number of PCR cycles followed by Southern blotting of the products so that the later nonquantitative part of the PCR reaction is avoided; and the use of low PCR cycle numbers using fluorescently labelled primers or incorporating fluorescent nucleotides.

It is wise to confirm any deletion detected in this manner with a different technique such as Southern blotting, as a base polymorphism at the 3′ end of a primer can lead to a false positive result.

3.3 PCR detection of trinucleotide repeat expansions

The process by which simple tandem repeats increase or decrease in length has been termed 'dynamic mutation' (Richards and Sutherland 1992). Dynamic mutation of 3-bp (trinucleotide or triplet) repeats within genes has been a surprising cause of neurological and degenerative genetic disease. In all cases, the disease is caused by expansion of a trinucleotide repeat either in the coding region of the gene, the 5′ or 3′ untranslated region or in an intron. At the time of writing, such diseases have only been identified in humans. The diseases caused range from the dominantly inherited Huntingdon's disease to the recessively inherited Friedreich's ataxia. Trinucleotide repeat

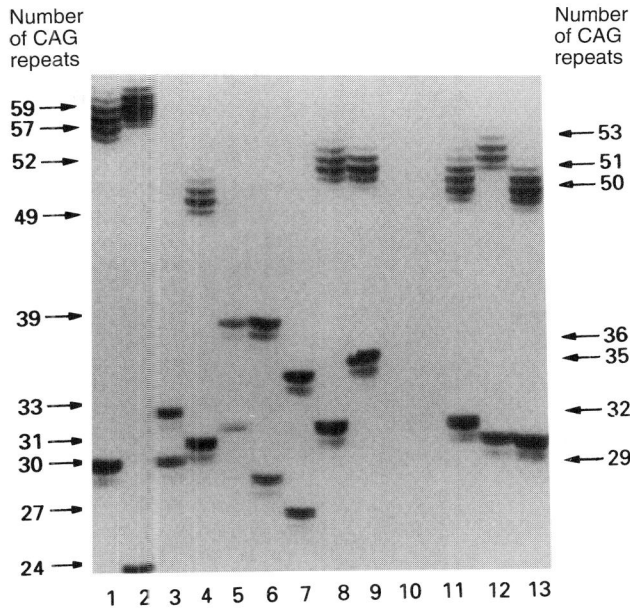

Fig 3.5

PCR analysis of the CAG repeat in the human *SCA1* gene. Expanded alleles with greater than about 40 CAG repeats are disease-causing. Patients' samples are those in lanes 1, 2, 4, 8, 9, 11, 12, and 13. (Reproduced with permission from Castellvi-Bel *et al.* 1994; BMJ Publishing Group.)

expansions have also been suggested to be involved in the aetiology of chromosomal rearrangements and in cancer pathogenesis.

Although the trinucleotide repeats involved in these diseases are polymorphic in length in the normal population, diseased individuals have alleles sized above a particular threshold. In detecting expansion mutations, it is therefore important to accurately size the alleles. This analysis is usually performed by PCR and the products sized on agarose or acrylamide gels. Problems associated with the analysis include difficulty in amplifying the longer alleles and difficulty in amplifying GC-rich repeats such as CGG. The replacement of dGTP with 7-deaza-2′-dGTP and the inclusion of DMSO or formamide in the PCR reaction ameliorate both of these problems. The yield of the amplification is not always good and ethidium bromide staining is weak when 7-deaza-2′-dGTP is used, so products are usually detected by either the inclusion of an isotope in the reaction, or the blotting of the gel followed by probing with a radiolabelled probe or a hapten-labelled probe which is detected by chemiluminescence.

Further reading

Buckle, V. J. and Kearney, L. (1994). New methods in cytogenetics. *Current Opinion in Genetics and Development*, **4**, 374. A review of state-of-the-art techniques in cytogenetics.

Chamberlain, J. S., Gibbs, R. A., Ranier, J. E., and Caskey, C. T. (1991). Detection of gene deletions using multiplex polymerase chain reactions. In *Methods in molecular biology*, Vol. 9 (ed. C. G. Mathew), pp. 299–312, Humana Press, Clifton, NJ. Provides detail of the methodology for multiplex PCR.

Darling, D. C. and Brickell, P. M. (1994). *Nucleic acid blotting: the basics*, IRL Press at Oxford University Press, Oxford. Provides details of Southern and Northern blotting.

Sutherland, G. R. and Richards, R. I. (1995). Simple tandem DNA repeats and human genetic disease. *Proceedings of the National Academy of Sciences, USA*, **92**, 3636. A review of trinucleotide repeat diseases and mutational mechanisms.

References

Castellvi-Bel, S., Matilla, T., Banchs, M. I., Kruyer, H., Corral, J., Milà, M., *et al.* (1994). Chemiluminescent detection of blotted PCR products (CB-PCR) of two CAG dynamic mutations (Huntingdon's disease and spinocerebellar ataxia type I). *Journal of Medical Genetics*, **31**, 654.

Gibbs, R. A., Nguyen, P.-H., Edwards, E., Civitello, A. B., and Caskey, C. T. (1990). Multiplex DNA deletion detection and exon sequencing of the hypoxanthine phosphoribosyltransferase gene in Lesch–Nyhan families. *Genomics*, **7**, 235.

Richards, R. I. and Sutherland, G. R. (1992). Dynamic mutations: A new class of mutations causing human disease. *Cell*, **70**, 709.

Schwartz, D. C. and Cantor, C. R. (1984). Separation of yeast chromosome-sized DNAs by pulsed field gradient gel electrophoresis. *Cell*, **37**, 67.

Vollrath, D. and Davis, R. W. (1987). Resolution of DNA molecules greater than 5 megabases by contour-clampled homogeneous electric fields. *Nucleic Acids Research*, **15**, 7865.

4 Diagnostic mutation detection methods

The methods to detect unknown mutations (scanning methods) are relatively difficult to perform, and so these methods are generally not used to detect mutations in cases where the mutation is known. In these cases the simpler diagnostic methods are used. Examples of such cases might be:

- when the mutation in one affected individual in a family has been characterized, and the other family members need to be diagnosed as being carriers of the mutation or not
- when a common disease-causing mutation is present at a high level in the general population, e.g. the ΔF508 cystic fibrosis mutation
- when a variant identified by a mutation scanning method needs to be evaluated as being a mutation or a polymorphism
- when you are scoring polymorphic variants

In this chapter, six of the most commonly used diagnostic methods are described. Allele-specific PCR (Section 2) and the oligonucleotide ligation assay (Section 3), diagnose the variant by the success or failure of amplification. Allele-specific oligonucleotide hybridization (Section 1), primer-introduced restriction analysis (Section 4), mini-sequencing (Section 5), and the 5' nuclease assay (Section 6) score for the presence of the mutation in the amplification product.

Each method has relative merits and draw-backs. Once you have understood what each method has to offer, you should decide what are the most important considerations for the project you are intending to do. For example, you might want to avoid using radioisotopes; you might be working on a very limited budget and therefore need to choose a cheap method; you might be needing to process a very large number of samples and therefore be seeking a method amenable to automation. At the end of this chapter the attributes of each of the different methods will be compared.

1. Allele-specific oligonucleotide hybridization

◇ Advantages:
- simple
- cheap
- gel-free

Under specific hybridization conditions, an oligonucleotide will only bind to a PCR product if the two are fully matched. A single base pair mismatch is sufficient to prevent hybridization. The use of a pair of

◇ Disadvantages:
● Precise hybridization conditions required

oligonucleotides, one carrying the mutant base and the other carrying the wild type base can be used to determine a PCR product as being homozygous wild type, heterozygous wild type/mutant, or homozygous mutant for a particular known mutation. This is termed allele-specific oligonucleotide hybridization or the 'dot-blot'.

Conventionally the PCR product is fixed onto a nylon membrane and probed with a labelled oligonucleotide. In the 'reverse dot-blot', an oligonucleotide is fixed to a membrane and probed with a labelled PCR product.

Traditionally, the probes have been isotopically labelled, but the method is increasingly being adapted to nonisotopic labelling methods.

1.1 Introduction

In an allele-specific oligonucleotide (ASO) hybridization or 'dot-blot', a number of PCR amplified samples can be typed for a single known mutation (Wallace *et al.* 1979; Saiki *et al.* 1986). The PCR amplified DNA is denatured and spotted onto a nylon membrane using a vacuum dot blotter. A labelled oligonucleotide probe (homologous to the wild-type or mutant sequence) is then hybridized to the membrane. The probe should be 19–21 bases long. As with PCR primers, self-complementarity should be avoided whenever possible. The potentially mismatched base, should be situated in the middle of the oligonucleotide, as this is the least stable position for a base pair mismatch.

◇ The hybridization conditions are governed by both the length of the oligonucleotide and GC-content.

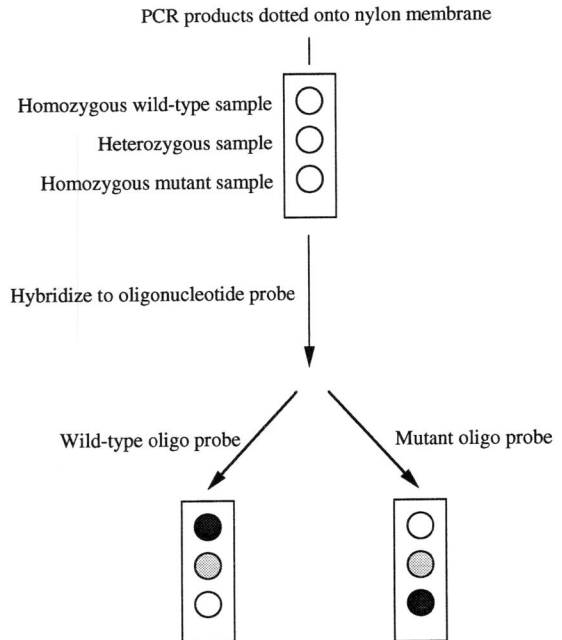

Fig 4.1

Genotyping individuals for a known mutation by allele-specific oligonucleotide hybridization.

The experimental difficulty of the method is in establishing the correct hybridization and wash stringencies. High stringency of the post hybridization washes, as produced by high temperature or low salt concentration results in the washing off of mismatched oligonucleotide whilst maintaining hybridization of the fully matched oligonucleotide (*Figure 4.1*). Thus the dot-blot is a simple method for the scoring of a small number of known mutations in a large number of samples. Insertions, deletions and base substitutions can all be detected.

The limitation of the technique is that only one mutation or polymorphism can be typed in a pair of hybridization reactions. In situations where it is desirable to type several or many mutations or polymorphisms in a relatively small number of samples, the reverse dot-blot becomes a viable alternative, e.g. in routine screening for the major cystic fibrosis and thalassaemia mutations. The reverse dot-blot (sometimes called the 'blot-dot') is so called because oligonucleotides for the common mutations and their wild type counterparts are spotted onto the membrane and probed with a labelled PCR product (or products) from a single individual (*Figure 4.2*). The production of a reverse dot-blot, however, is not as simple as the ordinary dot-blot as all oligonucleotides must hybridize with equal efficiency, so that each is sequence-specific under a single hybridization condition.

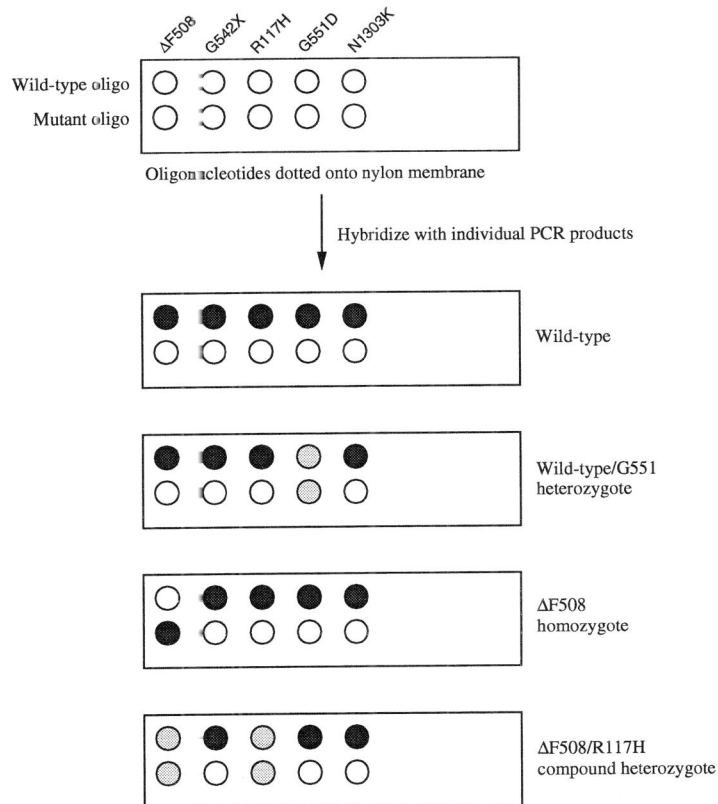

Fig 4.2

Simultaneous genotyping of the major cystic fibrosis mutations using the reverse dot blot allele specific hybridization.

1.2 The dot-blot methodology

1.2.1 Attachment of amplified DNA to the membrane

The DNA of interest should be PCR amplified in a conventional manner. The PCR products should be roughly quantitated, so that approximately equal amounts of each product are loaded onto the membrane. This should be done by running an aliquot of the PCR product on an agarose gel, so that the amounts can be normalized, i.e. if one sample has amplified less well than the others, correspondingly more of that sample should be blotted than the others. PCR product (100–200 ng) should be denatured in 100 μl of 0.4 M NaOH, 25 mM EDTA for 10 minutes at room temperature.

A piece of 3MM paper (Whatman) and a piece of charged nylon membrane (e.g. Hybond N+; Amersham), should be cut to the size of a dot blotter (e.g. Schleicher & Schuell Minifold I or Bio-Rad BioDot). A corner of the membrane should be snipped off, as an orientation marker. The 3MM paper and membrane should be pre-wetted in 2× SSC and placed in the dot-blotter with the membrane upper-most. A tap-driven suction line should be applied to the dot-blotter. Without much time delay after applying the suction, the DNA samples should be applied to the dot-blotter, one sample per slot. The liquid will be sucked through rapidly, leaving the DNA trapped on the membrane. The dot-blotter should be left as it is for 10 minutes allowing, to some extent, the NaOH to fix the DNA to the membrane. The blotter should then be opened up and the filter rinsed in 2× SSC. It is advisable in initial experiments to ensure complete fixing by baking the membrane at 80 °C for 1 hour or by UV cross-linking.

The dot blotter should be thoroughly cleaned after use, to avoid contamination from one experiment to the next.

◇ It is possible to check that the DNA samples have not leaked from slot to slot by adding bromophenol blue to the sample at 0.0001 per cent.

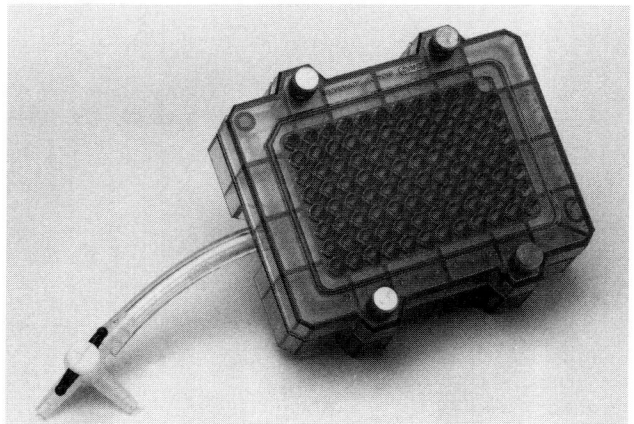

Fig 4.3

A dot blotter unit with vacuum line.

◇ For an oligonucleotide of mol. wt. 5600, 10 pmoles = 5600/100 = 56 ng.

◇ The mol. wt. can be crudely calculated by multiplying the number of bases in the oligonucleotide by 330, e.g. for a 20 mer, 20 × 330 = 6600.

◇ The exact molecular weight of the oligonucleotide can be calculated using the formula: $(249 \times nA) + (240 \times nT) + (265 \times nG) + (225 \times nC) + (63 \times n - 1) + 2$; where n is the number of bases of that type.

◇ The molarity of an oligonucleotide can be calculated as: concentration (mg/ml)/mol. wt. × 10^6 = μMolar. The number of mmoles can be calculated as: No. of mg/mol. wt.

◇ The formula: $T_m = 2\ °C$ (number of A + T bases) + 4 °C (number of G + C bases), provides a crude and very approximate guide to the T_m of the oligonucleotide, e.g. a 20 mer consisting of 11 G + Cs and 9 A + Ts, will have a T_m in the region of 62 °C.

1.2.2 Labelling the oligonucleotide

Conventional oligonucleotide synthesis leaves the 5' end of the molecule unphosphorylated. This permits labelling by T4 polynucleotide kinase, which transfers the labelled phosphate of a [γ-^{32}P]ATP molecule onto the 5' end of the oligonucleotide.

Many workers choose to determine the efficiency of labelling and to purify the labelled oligonucleotide away from free label. In most cases neither is necessary. Successful labelling of the oligonucleotide can be determined by the inclusion of a positive control on the membrane, i.e. a PCR product fully homologous to the oligonucleotide. Only if a background is present on the membrane after hybridization, is it necessary to purify the probe. Purification of the free label away from the labelled oligonucleotide, does however serve to make the probe less 'hot' and therefore better for handling in the hybridization and washes.

1.2.3 Hybridization (in ordinary buffer)

Hybridization is carried out as for Southern blotting. Membranes should be prehybridized for 10–30 minutes. Hybridization can be carried out overnight, but periods as short as 1 hour should be sufficient. It is not necessary to boil the probe prior to hybridization. The temperature of hybridization should be approximately 10 °C lower than the estimated T_m. The post-hybridization washes should initially be carried out at the hybridization temperature or lower. The inclusion of positive controls and negative controls (i.e. fully matched and mismatched sequences, respectively) is a great aid in the washing of the membrane. As you wash the filter, you should monitor the positive and negative control regions with a Geiger-counter. If both are hot, you need to increase the wash temperature by 2 °C. Then check the membrane again. You should keep doing this until the signal from the negative control has dropped off, whilst maintaining the signal from the positive control. The membrane is then ready for exposure to film. Following exposure, filters can usually be stripped for probing with the second oligonucleotide. If the filter fails to strip completely, a duplicate membrane should be made.

Fig 4.4

Diagnosis of a heterozygous point mutation.

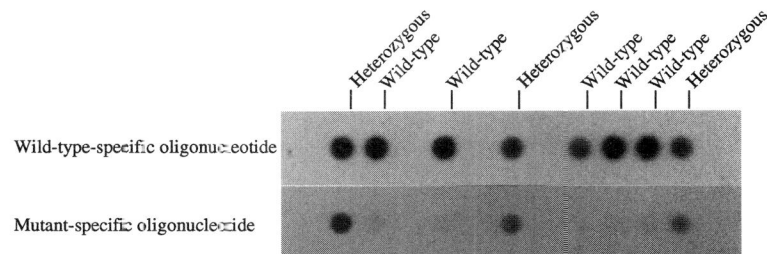

1.2.4 Hybridization in TMAC buffer

The inclusion of tetramethyl ammonium chloride (TMAC) in the hybridization buffer eliminates the need to consider the GC content of the oligonucleotide. TMAC reduces hydrogen bonding energies between G–C base pairs, whilst binding to A–T base pairs, increasing their hydrogen bond thermal stabilities. TMAC is added to the hybridization buffer to a concentration of 3 M; at which the differences between the stabilities of G–C versus A–T base pairs are minimized. For 20-mer oligonucleotides, hybridization should be carried out at approximately 56 °C and post-hybridization washes at about 60 °C. Beware that the probe will dissociate over a smaller temperature range in TMAC buffer than in ordinary hybridization buffer. Finally the membrane should be rinsed in 2× SSC or 2× SSPE.

The use of TMAC buffer also permits more than one oligonucleotide probe to be used in a single hybridization reaction, provided the oligonucleotides are all of the same length.

TMAC solid is highly hygroscopic and so the molecular weight will alter as it accumulates moisture. Thus when making a solution from the solid it is necessary to establish the true concentration using a refractometer and the formula:

$$\text{Molar concentration} = (N - 1.331)/0.018$$

where N is the refractive index. However, TMAC is obtainable (at considerable cost) as a pre-made 5 M solution from Sigma.

1.3   Modifications of ASO hybridization

1.3.1 The reverse dot-blot

The reverse dot-blot allows the simultaneous typing of several known mutations in a PCR product or products from a single individual. In this variant of the ASO method, the oligonucleotides are blotted onto the membrane and are probed with PCR product (Saiki *et al.* 1989; Wall *et al.* 1995).

In order to fix a sufficient quantity of an oligonucleotide to a membrane, it is necessary to tail the oligonucleotide with thymidine bases. This is done using the enzyme terminal deoxyribonucleotidyltransferase. The size of the poly d(T) tail added to the oligonucleotides is not critical, but should be about 400 bases. The same tailing conditions should be used for each oligonucleotide, so that each has approximately the same tail length.

The tailed oligonucleotide is then UV co-linked to the membrane, leaving the specific sequence available for hybridization to the PCR product probe. The PCR products can be labelled by either performing the amplification with a kinase-labelled primer (see Chapter 5, Section 3.3.1) or by kinase labelling the PCR product.

In order to get even hybridization of each oligonucleotide to its homologous sequence under a single hybridization, it is necessary to conduct the hybridization in TMAC so the effect of base composition can be eliminated. It may also be necessary to adjust the levels of individual oligonucleotides on the membrane. This is because secondary structures may exist in some parts of the PCR product probe, or if the membrane is being probed with more than one PCR product at a time, secondary structures may exist in one of the probes. Any secondary structure will reduce hybridization and therefore signal. The level of signal can be increased to a level similar to that of the other oligonucleotides on the membrane by simply increasing the quantity of the oligonucleotide in question.

1.3.2 Multiplex ASO analysis

If several mutations within a gene are to be typed, it may be possible to perform a multiplexed ASO experiment, rather than the reverse dot-blot (Seradheera *et al.* 1995). A prerequisite is that all oligonucleotides to be handled together must have very similar melting behaviours. Under well tried and tested conditions it may be possible to probe a membrane with a pool of mutant oligonucleotides and subsequently investigate those samples which prove positive.

1.3.3 Nonisotopic labelling

There are a variety of methods for the nonisotopic labelling of DNA, but only a few have the sensitivity to rival isotopic labelling. In techniques such as allele-specific oligonucleotide hybridization, high sensitivity is not required as the target is PCR amplified and therefore present in a large quantity (Bugawan *et al.* 1990).

There are essentially two types of method for nonisotopic detection: chemiluminescence and chromogenic detection. Chemiluminescence produces light, which is detectable by an imager or exposure to X-ray film. Chromogenic methods produce a coloured precipitate directly on the membrane. The chromogenic methods are not as sensitive as the luminescence methods. But as high sensitivity is not required, chromogenic methods are the usual form of nonisotopic ASO hybridization.

Oligonucleotides may be labelled with enzymes such as horseradish peroxidase (HRP) or alkaline phosphatase (AP). Following ASO hybridization, the bound probe is detected by incubation with an appropriate chromogenic substrate. Biotinylated oligonucleotides can be detected via a streptavidin–HRP conjugate. In this case, an HRP-labelled antibody is conjugated to streptavidin, which stably binds to biotin. Primers may also be labelled with the steroid hapten digoxigenin (DIG) and detected via an anti-DIG-AP conjugate in an enzyme-linked immunoassay.

Labelled oligonucleotides can be purchased from specialized oligonucleotide synthesis companies. But beware, enzyme-labelled oligonucleotides are especially expensive.

2. Allele-specific PCR

Most of the mutation diagnostic methods and scanning methods are procedures which analyse PCR-amplified DNA. The allele-specific polymerase chain reaction differs in that the assay occurs within the PCR reaction itself. Sequence-specific PCR primers which differ from each other at their terminal 3' nucleotide are used to amplify only the normal allele in one reaction, and only the mutant allele in another reaction. Amplification is scored by agarose gel electrophoresis. The method can be multiplexed to permit simultaneous analysis of several known mutations.

The appeal of this detection method is that it is relatively simple and requires no specialist or expensive equipment.

2.1 Introduction

Allele-specific PCR, often called the amplification refractory mutation system (ARMS), is a simple method for the detection of any known point mutation or small deletion or insertion (Newton *et al.* 1989; Wu *et al.* 1989). It utilizes three primers for two separate and complementary PCR amplifications. One primer is common to the two reactions and the other two primers, which differ from each other at their 3' ends, are specific for the particular variant base in question.

When the 3' end of a specific primer is fully matched, amplification occurs. When the 3' end of a specific primer is mismatched, amplification fails to occur. The PCR products are scored on an agarose gel. The genotype of a (homozygous) wild-type sample is characterized by an amplification product only in the normal reaction. A heterozygote is characterized by amplification products in both

Wild-type specific reaction

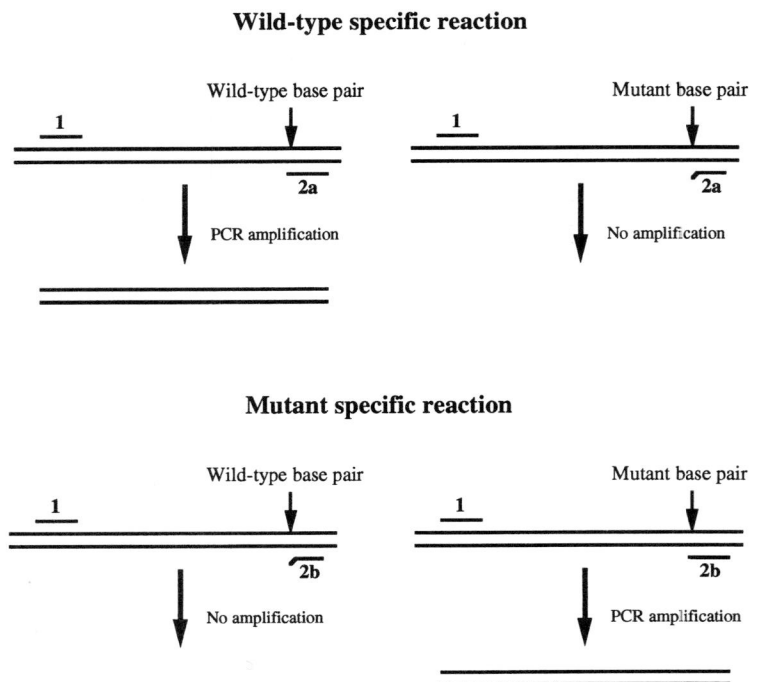

Mutant specific reaction

Fig 4.5

The basis of allele-specific PCR

reactions, and a homozygous mutant sample generates product in only the mutant reaction.

The allele-specific primers are sufficiently long so that each will hybridize equally well to each allele without permitting an appreciable level of nonspecific mispriming. Thus the specificity of the reaction is controlled not by the annealing temperature but by the ability of *Taq* polymerase not to extend a primer mismatched at the 3′ end.

2.1.1 Primer design considerations

Despite the PCR primer having a mismatch at the 3′ end, *Taq* polymerase will actually extend the primer, albeit at a very much lower rate than normal. The ability of *Taq* polymerase to extend a 3′ terminally mismatched primer, probably depends not only upon the strength of the non-Watson–Crick base pair, but also on the level of stacking forces (see Chapter 5, Section 4.1) and overall steric structure of the terminus to be extended.

In order to ensure that the mismatched primer is not extended at all (or not detectably) a further mismatch may be introduced to destabilize it further (Newton *et al.* 1989; Kwok *et al.* 1990). This secondary mismatch is usually at the position of the 3′ penultimate base.

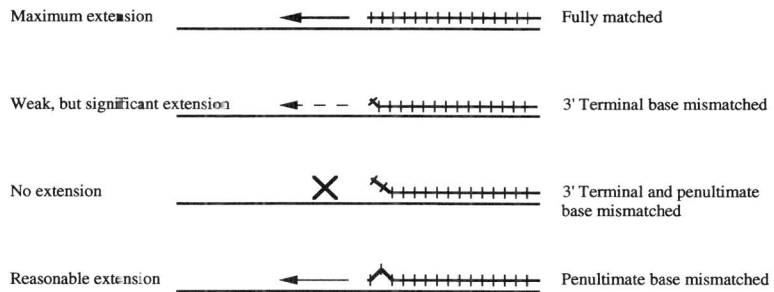

Fig 4.6

The inclusion of a secondary mismatch in the PCR primer improves specificity.

The degree of destabilization should be sufficient to prevent extension of the primer mismatched at its 3′ terminal base but not so strong as to prevent extension of the primer matched at its terminal base. Additionally the degree of destabilization required depends upon the nature of the 3′ terminal mismatch. Purine/pyrimidine mismatches form relatively strong non-Watson–Crick base pairs, whereas purine/purine or pyrimidine/pyrimidine mismatches form very weak pairs (see Table 4.1).

Table 4.1 Relative strength of Watson–Crick versus non-Watson–Crick base pairs

Base pair	Base pairing strength	Destabilizing strength
G–C, A–T	Maximum	None
G–T, A–C	Strong	Weak
G–G, A–A	Medium	Medium
C–C	Weak	Strong
G–A, C–T, T–T	Very weak	Maximum

In order to destabilize a strongly pairing 3′ terminal mismatch it is necessary to have a relatively strong destabilizing secondary mismatch. For example a T–G (strongly base-pairing/weakly destabilizing) 3′ terminal mismatch would be extended relatively efficiently by *Taq* polymerase and thus requires a strong secondary destabilizing mismatch. If the penultimate base on the target DNA strand were a 'G', a G–A mismatch (strongly destabilizing/weakly base-pairing) would be required to provide sufficient destabilization (see Table 4.2).

2.1.2 Multiplex analysis

There may be several common mutations in your gene of interest. Separate allele-specific PCR amplifications can be performed for each mutation, but much time and effort could be saved by performing multiplex PCRs (Ferrie *et al.* 1992). This involves the simultaneous amplification of more than one target in a single reaction tube. The difficulty in doing this is getting all the primers involved to work under a single condition and without them interfering with each other.

The allele-specific PCR primers are large so that they will hybridize equally well to the wild-type and mutant alleles. This also allows the window of annealing temperature to be relatively large, enabling primer pairs of moderately different melting temperatures to be simultaneously used for amplification under a single condition without background nonspecific amplification becoming much of a problem. The positioning and design of the PCR primers will depend upon the nature and spacing of the mutations to be scored. It is necessary that the PCR products are of different sizes so they can be resolved on the agarose gel. If for example, three widely spaced mutations are to be multiplexed, the primers might be spaced to give 200 bp, 250 bp, and 300 bp sized products. If two or more closely spaced mutations are to be scored, it may be possible to use a single (non-allele-specific) primer common to each product.

A problem in allele-specific PCR arises when two mutations lie within a few base pairs. In such a case, primer sets specific to each mutation may be needed and separate reactions performed.

Table 4.2 Table of penultimate destabilizing mismatches appropriate for particular terminal mismatches[a]

3′ terminal mismatch	3′ penultimate destabilizing mismatch			
	G	A	T	C
T–T	T	C	G	A
T–C	T	C	G	A
T–G	A	G	C/T	T
A–A	G	A	G	A
A–C	A	G	T	C
A–G	T	C	G	A
C–C	T	C	G	A
G–G	G	A	G	A

[a] Data taken from Little (1994).

2.2 Methodology

As this technique consists solely of PCR amplification many of the technical considerations are the same as those for PCR. The primers are generally large, about 30 bases in length and the GC-content should be about 50 per cent. The usual considerations of primer design apply, i.e. avoidance of repeats, palindromes, and complementarity between primer pairs. The amplifications should be optimized to give a clean product with little or no background of products generated by mispriming. This can be controlled by adjustment of the concentrations of magnesium chloride, primer, and enzyme. The type of buffer used, annealing temperature, and hot starts may also minimize background. Cycle numbers in excess of 35 may generate background. It is important that a polymerase lacking a 3′ to 5′ exonuclease proof reading activity is used, such as *Taq*—but not *Vent* or *Pfu,* as the maintenance of mismatches is essential.

Primers should be positioned far enough apart to be easily resolvable on agarose gels i.e. >200 bp. If the PCR is to be multiplexed, products should be designed to differ in length by ≥50 bp.

The PCR products should be run on agarose gels of an appropriate concentration for the resolution of the fragment size(s) produced. The resolution of fragments below 250 bp can be improved by the use of 'Nusieve agarose (FMC BioProducts). This agarose though can misbehave in the microwave due to superheating and produce brittle, easily broken gels. A good compromise is to mix ordinary and Nusieve agarose to combined concentrations of up to 3 per cent. The gel and running buffer should contain ethidium bromide at 0.5 μg/ml. This stains the DNA by intercalation and allows visualization by glowing when the gel is placed on a UV transilluminator. A photograph should be taken to provide a record of the gel.

2.2.1 Dual colour labelling

The use of ethidium bromide stained agarose gels requires pairwise reactions to be performed (wild-type- and mutant-specific) as the wild-type and mutant products are indistinguishable. If the 5′ ends of the allele-specific primers were labelled with different fluorescent labels, and the 5′ end of the common primer biotin labelled, the wild-type-specific and the mutant-specific reactions could be performed in a single tube (in much the same manner as in the oligonucleotide ligation assay, see Section 3). The advantages of this approach are that gel electrophoresis is not required and that the method is amenable to automation. The disadvantage of the labelled approach is that this 'low-tech' simple technique now requires expensive solid phase and fluorescent detection technology.

3.   Oligonucleotide ligation assay

When two oligonucleotides, annealed to a strand of DNA are exactly juxtaposed, they can be joined by the enzyme DNA ligase. A single base pair mismatch at the junction of the two oligonucleotides, is

◇ Ethidium bromide is a mutagen and a probable carcinogen. It must therefore be handled with care and gloves should always be worn. Any spillages should be thoroughly cleaned up.

◇ Ultraviolet light is damaging to the eyes and skin. Eye protection or a transilluminator cover should always be used when visualizing gels.

◇ Advantages:
● Gel-free
● Nonisotopic
● Potential for high throughput

◇ Disadvantages:
● High throughput assay requires specialist equipment

sufficient to prevent ligation. Rather than assaying ligation by gel electrophoresis and visualization of a new larger sized DNA fragment, ligation is scored by assaying for labels on the two oligonucleotides becoming present on a single molecule.

When ligation is scored by ELISA and reactions are conducted in 96-well microtitre plates, the oligonucleotide ligation assay can be performed by robot and the results analysed by plate reader and fed directly into a computer. The method is therefore an excellent one for the scoring of a known mutation in a large number of samples.

The assay comes in two main forms: the oligonucleotide ligation assay, which is performed on PCR-amplified DNA; and the ligase chain reaction, which is performed on genomic DNA and amplified with a thermostable DNA ligase.

3.1 The theory

The oligonucleotide ligation assay (OLA) typically consists of a 5′ biotinylated left-hand oligonucleotide and a 3′ labelled right-hand oligonucleotide which is phosphorylated at the 5′ end. The label may be isotopic, but is more commonly nonisotopic. The two oligonucleotides are annealed to PCR-amplified target DNA. If the 3′ most base of the left-hand oligonucleotide is fully matched to the target DNA, when treated with DNA ligase, the left-hand biotinylated oligonucleotide will become covalently joined to the labelled right-hand oligonucleotide creating a molecule not previously present in the reaction mixture.

Fig 4.7

DNA ligase covalently links two immediately adjacent oligonucleotides on a target DNA molecule. If a mismatch is present at the junction, ligation fails to occur.

3′ END OF LEFT HAND OLIGO COMPLETELY MATCHED

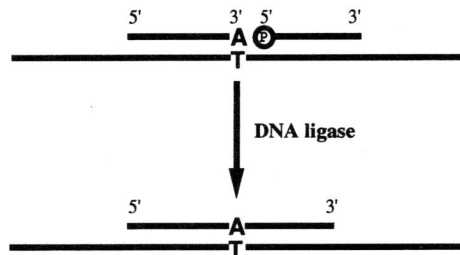

3′ END OF LEFT HAND OLIGO MISMATCHED

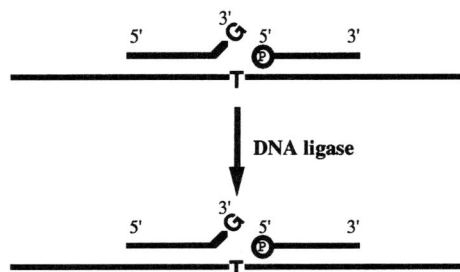

If the 3′ most base of the left-hand oligonucleotide is mismatched, ligation to the right-hand oligonucleotide is effectively prevented; proceding at a rate reduced approximately 100-fold (Landegren *et al.* 1988) (*Figure 4.7*).

Pairwise reactions are performed, one using a left-hand oligonucleotide in which the 3′ most base is complementary to the wild-type base in question and another using a left-hand oligonucleotide which is specific for the mutant base.

The ligase joins the DNA molecules by converting the energy of a pyrophosphate linkage in an ATP molecule into a phosphodiester bond between the hydroxyl group at the 3′ end of the left-hand oligonucleotide and the phosphate group at the 5′ end of the right-hand oligonucleotide.

The left-hand oligonucleotide is biotinylated so that it may be captured onto a streptavidin coated solid support. If ligation has occurred, the label at the 3′ end of the right-hand oligonucleotide will become linked to the biotinylated left-hand molecule. Hence, when captured onto a solid support, the label will go into the solid phase; generating a positive signal. If ligation has not occurred (due to a mismatch at the junction), the label will stay in the liquid phase and be washed off, thus giving a negative signal (*Figure 4.8*).

A wild-type DNA sample is characterized by a positive signal from the wild-type left-hand oligonucleotide and a negative signal from the

Fig 4.8

When ligation occurs, the label can be attached to a solid support via a Biotin–Streptavidin link.

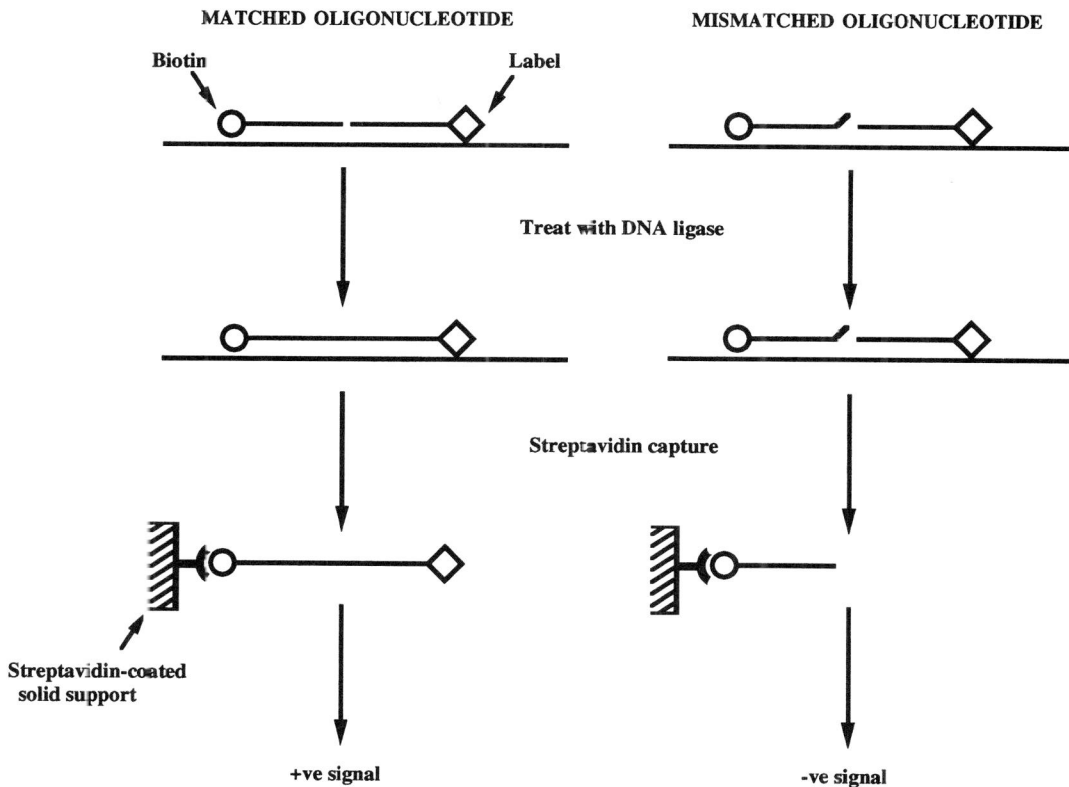

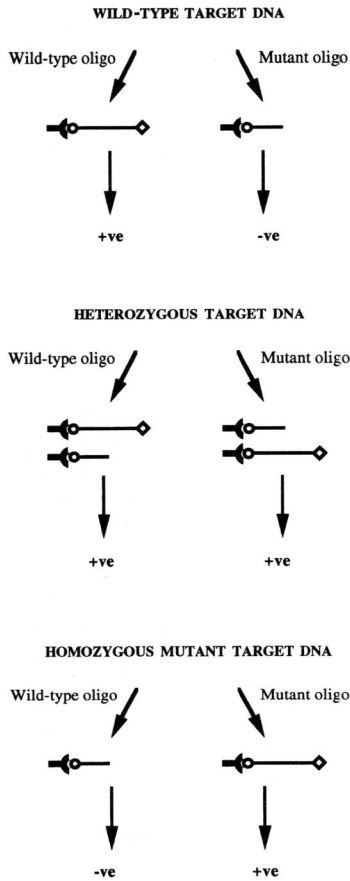

Fig 4.9

Oligonucleotide ligation assay diagnosis of homozygous and heterozygous samples.

mutant left-hand oligonucleotide (and vice versa for a homozygous mutant sample). A heterozygous DNA sample is characterized by positive signals from both the wild-type and mutant oligonucleotides.

3.1.1 Solid supports

When a molecule becomes attached to something solid, it can be said to be attached to a solid-support or be 'in the solid-phase'. The nature of the solid support in the OLA may be either streptavidin-coated microtitre plates, streptavidin-coated manifolds (which sit in the wells of microtitre plates), or streptavidin-coated magnetic beads. When magnetic beads are used, the beads and attached DNA molecules can be drawn to the edge of the tube or microtitre plate well using a strong magnet, whilst the liquid is washed away. Streptavidin-coated microtitre plates and manifolds allow the biotinylated molecules to become attached to the surface of the plastic without any external interference.

3.1.2 Dual colour assay

Fluorescent labels have found their way into OLA analysis as well as many of the other mutation detection techniques. If instead of the biotin group being added to the 5′ end of the left-hand oligonucleotide and the label added to the 3′ end of the right-hand oligonucleotide, the biotin group is added to the 3′ end of right-hand oligonucleotide and different fluorescent labels added to 5′ ends of the wild-type- and mutant-specific left-hand oligonucleotides, all three oligonucleotides can be included in a single reaction. This eliminates the need for separate wild-type- and mutant-specific reactions (Samiotaki *et al.* 1994).

When the sample is homozygous wild-type or homozygous mutant, one or other colour will be present in the resulting solid phase. When the sample is heterozygous, both colours will be detected. The dual colour design also provides an internal control, reducing the need for replicate analyses which are often required for reliable interpretation of conventional OLA experiments. The down side of the dual colour assay is that it requires a fluorometric plate reader, which is unlikely to be cheap.

3.1.3 OLA methodology

Each sample of PCR-amplified DNA is diluted and placed in a well of a 96-well microtitre plate. The samples are then heated to denature, and then allowed to cool to below 60 °C before the ligation mix (each oligonucleotide, buffer, ATP, and ligase) is added. The ligation reactions are incubated at 37 °C for 30 minutes.

The samples should then be transferred to washed and blocked streptavidin-coated microtitre plates. The samples are then incubated to bind streptavidin and washed as recommended by the plate manufacturer.

If the label is digoxigenin, ELISA detection reactions should be performed and spectrophotometric absorbances analysed by a plate reader. The results can be directly fed into a computer for storage (Nickerson *et al.* 1990).

If you do not have such equipment at your disposal and cannot afford streptavidin coated microtitre plates, the protocol can be modified so that the biotinylated molecules are captured onto magnetic beads. After washing away the liquid phase, ligation can be scored by dot-blotting and autoradiography if the label is isotopic or by ELISA and visual analysis.

3.2 Ligase chain reaction

The Ligase chain reaction (LCR) is a modification of the oligo-nucleotide ligation assay (Barany 1991). It has much greater sensitivity due to the amplification of ligated oligonucleotides through repeated cycles of denaturation, oligonucleotide annealing, and ligation using a thermostable ligase. Unlike PCR, LCR does not create new DNA, but converts short oligonucleotides into longer joined molecules.

In the LCR, six oligonucleotides are required: two for each strand of the target DNA and an additional two to recognize the mutant base. When fully matched, the adjacent annealed oligonucleotides are ligated to form continuous molecules as in the OLA. The ligated products are then melted away from the target DNA allowing both to become targets for further oligonucleotide annealing. With each cycle

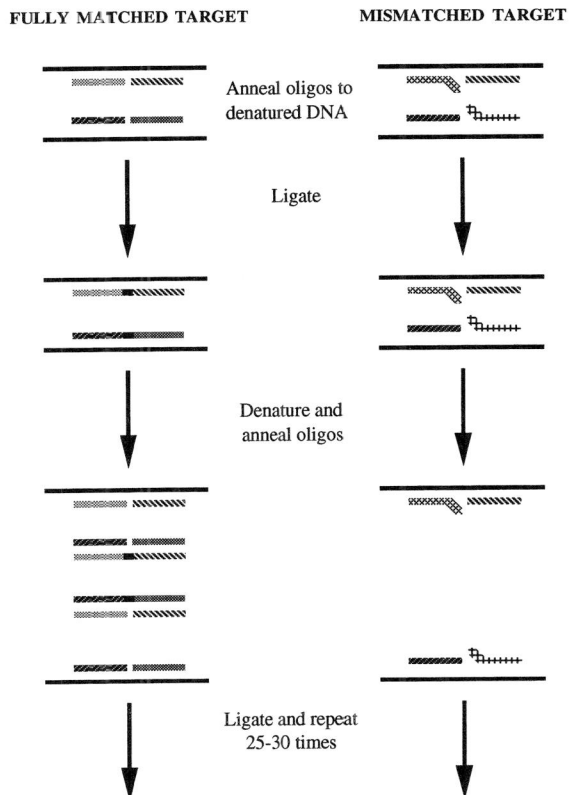

FULLY MATCHED TARGET **MISMATCHED TARGET**

Anneal oligos to denatured DNA

Ligate

Denature and anneal oligos

Ligate and repeat 25-30 times

Fig 4.10

The steps involved in the ligase chain reaction.

of denaturation, annealing, and ligation, the number of ligated molecules doubles, thus amplifying the signal.

The advantage of the LCR over PCR followed by OLA, is that amplification and mutation diagnosis is combined into a single step. The disadvantage is that the LCR has quirks and difficulties of its own. A major problem of the method is that template-independent ligation can occur and serve as a target in subsequent cycles potentially causing a false positive signal. The level of target-independent ligation is influenced mainly by the concentration of the oligonucleotides, but also by the enzyme concentration and cycle number.

3.2.1 LCR methodology

Target DNA is amplified with each oligonucleotide and *Taq* DNA ligase. Each cycle consists of a denaturing step at 94 °C for about 30 s, and an annealing/ligation step at about 62 °C for about 4 minutes; 25–30 cycles should be sufficient. Ligation is detected and scored in the same manner as in the OLA.

4. Primer-introduced restriction analysis

◇ Advantages:
● Very simple
● Nonradioactive
● Cheap

◇ Disadvantages:
● Not always feasible
● Requires gel electrophoresis

Primer-introduced restriction analysis is a technique which allows known mutations to be diagnosed by restriction digestion. By introducing a base change close to the position of a known mutation by a mismatch in the PCR primer, it is possible to create a restriction endonuclease recognition site that is diagnostic for the mutation. Following restriction digestion, the different sized wild-type versus mutant PCR fragments, are analysed by gel electrophoresis.

4.1 The principles

Occasionally mutations create or destroy restriction endonuclease target sites. This enables the simple diagnosis of the mutation by gel electrophoresis of restriction-digested PCR amplified DNA. The constant production of novel restriction enzymes means that the likelihood of a given mutation being amenable to restriction diagnosis is ever increasing. However the majority of mutations do not lie in such fortunate sequence surroundings. In these cases, mutation diagnostic methods, such as those described in this chapter, are required. A cheap and simple alternative method is that of 'Primer-induced restriction analysis' (PIRA) (Haliassos *et al.* 1989; Jacobson and Moskovits 1991; Jacobson 1992). In this method, a mismatch is incorporated into the primer, which is situated close to the mutation site. The combination of the altered base in the primer sequence and the base at the mutation site, has the effect of creating a new restriction target site. The approach may be used to create a new target site on either the wild-type allele or the mutant allele.

For example, the A to T mutation in the context of GGGA-CATATCGATT in the wild-type and GGGACTTATCGATT in the mutant does not create or destroy a restriction site. If the (reverse) primer contains a mismatch such that the base 3 bp 3′ to the mutant base is converted from a T to a G, the sequence GGGACTTAGC-GATT is created in the mutant sequence, which harbours a target site for the restriction enzyme *Dde*I (recognition sequence—CTNAG) (*Figure 4 11*). In such a situation, the homozygous wild-type form would be characterized by a single band of the full-length size. The homozygous mutant form is characterized by a single band of the reduced size, and the heterozygous form by bands of both sizes.

One problem with the PIRA method is the difficulty of deciding what change in the primer sequence will cause what restriction site in association with the mutation. This may be done visually, by inspecting the lists of alphabetized recognition sites found in some molecular biologicals catalogues, or by using computer-based DNA sequence analysis programs such as the University of Wisconsin GCG package. It is perhaps easiest to use the computer program and alter bases in the vicinity of the mutant base one at a time, until a new enzyme target site is created.

A draw back of this method is that the introduced enzyme recognition site lies close to the end of the PCR product such that the uncut and cut molecules will only differ in size by approximately the size of the primer. Therefore the fragment size and/or the gel matrix and its concentration also have to be taken into consideration. In general, the fragment size will need to be small (i.e. less than 200 bp) and the gel for the fragment analysis will need to be high percentage agarose, perhaps containing NuSieve agarose (see Section 2.2), or be polyacrylamide. A greater uncut/cut differential can be generated by choosing to create recognition sites for restriction enzymes which cut at a position away from the recognition sites, e.g. *Bsg*I, which cuts 14 and

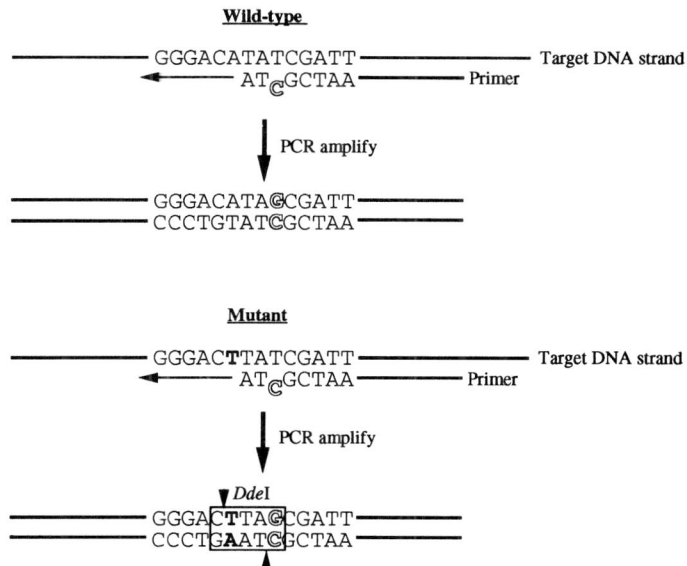

Fig 4.11

The PCR primer contains a mismatch which intoduces a change into the DNA. In conjunction with the mutation, a *Dde*I site is formed.

16 bp (on the different DNA strands) 3′ to the recognition sequence of 5′-GTGCAG-3′.

The limitation of the technique is that in some instances no change of sequence in the vicinity of the mutation will create a new target site. On the plus side, the technique is cheap and easy and is applicable to small deletions and insertions as well as point mutations.

4.2 Methodology

The annealing temperature of the PCR should be low enough to tolerate the mismatch in one of the primers yet high enough to maintain specificity. The usual annealing temperature range of 55–60 °C ought to be fine.

One quarter of a 20–50 μl PCR reaction ought to be ample for the restriction analysis. Problems may be encountered in getting the DNA to cut properly because of inhibition by PCR reaction buffer components or free nucleotides. It is therefore advisable, at least in initial experiments, to clean the DNA by ethanol precipitation. The use of ammonium acetate at 2 M as the salt in the precipitation inhibits the pelleting of free nucleotides. Alternatively, a centrifugation ultrafiltration unit such as a Centricon unit (Amicon) can be used to purify the amplified DNA away from salts, primers, and nucleotides.

The DNA should then be digested to completion and run on a suitable gel. DNA is visualized by ethidium bromide staining and UV illumination (see Section 2.2). Alternatively, if the gel is polyacrylamide, DNA may be visualized by silver staining (see Chapter 5, Section 5.5).

5. Mini-sequencing

◇ Advantages:
● Gel-free

◇ Disadvantages:
● Radioactive

Mini-sequencing is in effect a sequencing reaction, but in which only one base is added to the primer. Unlike conventional sequencing, the products are analysed by measuring label incorporation rather than by gel electrophoresis.

In the mini-sequencing reaction, a primer is annealed to a PCR product, immediately adjacent to the position of a known mutation. In two separate reactions, a labelled base complementary to the mutant base, and another complementary to the wild-type base are extended onto the primer by a DNA polymerase. Homozygotes are distinguished by low incorporation in one reaction and high incorporation in the other. Heterozygotes are characterized by high incorporations in both reactions.

The technique is amenable to modification for nonisotopic and robotic analysis and is sensitive enough to detect low levels of mosaicism.

5.1 Principles

The technique of mini-sequencing (or sometimes known as single nucleotide primer extension, or 'Snupe', in the world of quantitative PCR) can be used to diagnose any known point mutation, deletion, or insertion.

Identifying known mutations by DNA sequencing, might seem an unnecessarily complicated approach. Obtaining the sequence information at just a single base pair however, only requires the sequencing of that particular base. This can be done by including only one base in the sequencing reaction rather than all four. When this base is labelled (usually with ^{3}H) and complementary to the first base immediately 3' to the primer (on the target strand), the label will become incorporated onto the primer. If the base included in the reaction is not complementary to the base 3' to the primer, the label will not be incorporated. Thus a given base pair can be sequenced on the basis of label incorporation or failure of incorporation without the need for electrophoretic size separation (Syvänen *et al.* 1990). Two separate reactions are performed, one in which the nucleotide used is specific for the wild-type sequence and the other in which the nucleotide is specific for the mutant sequence.

Incorporation is assayed with the aid of biotin–streptavidin technology (see Section 3.1.1), hence the technique is often termed solid-phase mini-sequencing. The most common choice of solid support is the streptavidin-coated microtitre plate. However, streptavidin-coated magnetic beads, can also be used.

One of the primers used for the PCR amplification of the target is 5' biotinylated, so that the PCR product can be captured onto a solid support via a biotin–streptavidin link. The unbiotinylated DNA strand is then removed by alkaline denaturation and washing. The primer for the mini-sequencing reaction is then annealed to the captured strand. Following the extension reaction, the support is washed to remove unincorporated nucleotide. If the primer has been extended, the solid-phase will have become radioactive. Incorporation is measured by alkaline release of the primer and counting in a scintillation counter.

In the case of a 'G' to 'T' point mutation, for example, the complementary base on the antisense strand would be a 'C' in the wild-type and an 'A' in the mutant. If the captured strand is the antisense strand, the base included in the mutant-specific reaction would be a labelled 'T' (*Figure 4.12*).

Reactions which should not generate any counts, i.e. the mutant-specific reaction in a wild-type sample, always give a low level of incorporation rather than a zero level of incorporation. This may be due to a low level of misincorporation by the polymerase or failure to wash away all the unincorporated nucleotide. Thus a homozygote is characterized by a strong signal in one reaction and a weak signal in the other. A heterozygote is characterized by strong signals in both reactions. The results are presented as a ratio of the mutant-specific and the wild-type-specific reactions. This has the effect of eliminating sample to sample variation caused by variable efficiency of PCR

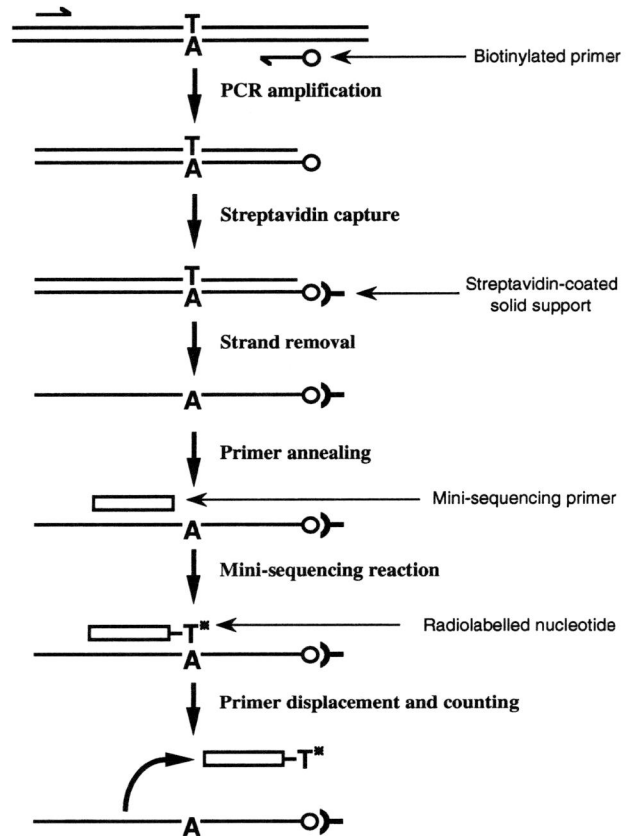

Fig 4.12

The solid-phase mini-sequencing procedure.

amplification. Typical results for a wild-type homozygote sample would be a wild-type-specific counts/mutant-specific counts ratio of approximately 100. A homozygous mutant sample would be expected to give a ratio of approximately 0.01. A heterozygous sample should yield a ratio of approximately 1.

The use of microtitre plates facilitates the analysis of large numbers of samples and potentially enables much of the work to be performed by robot (Ihalainen *et al.* 1994). The cost of the process is not as cheap as alternative methods such as allele-specific oligonucleotide hybridization or allele-specific PCR due to the cost of the streptavidin-coated microtitre plates.

5.1.1 Methodology

One of the primers for the PCR amplification of the DNA samples should be 5′ biotinylated and used at a final concentration of no more than 0.2 μM. Higher concentrations of biotinylated primer may result in a large quantity of unused primer remaining after the amplification, which will compete with the product for later streptavidin capture. Following amplification, approximately a tenth of the reaction volume is mixed with the capture buffer and transferred to a well of a streptavidin-coated microtitre plate. The plate should then be incubated at 37 °C with gentle shaking for 1.5 hours to allow solid-phase capture to take place. The wells are then washed to remove unbound material. The bound material is next denatured with sodium hydroxide, and the unbound strand removed by washing. The remaining bound single-stranded DNA is then ready for the mini-sequencing reaction. The enzyme used is typically *Taq* polymerase, as this enables the annealing and extension steps to occur simultaneously, i.e. at 50 °C or more. Following the mini-sequencing reaction, the plate is washed again, this time to remove unincorporated label. Finally, the primer is released by incubation with sodium hydroxide. The free primer is then transferred to a scintillation counter for analysis.

5.1.2 Detection methods

In most examples of solid-phase mini-sequencing the label used has been isotopic (Jalanko *et al.* 1992; Syvänen *et al.* 1992). Sufficient sensitivity is provided by ^3H, a weak β-emitter, with a relatively long half-life. Although care should be taken not to spill any of the labelled material and to dispose of the contaminated plastics correctly, it is not necessary to work behind protective shielding. Thus the need for alternative labels is not as strong in this method as it is in some others. Colorimetric and chemiluminescent detection methods are applicable, but require added steps (Harju *et al.* 1993) Fluorescence technology is also suitable: results could be read directly by a fluorometric plate reader or by electrophoretic analysis on automated sequencing gels.

The quantification of isotopic labels by counting the primer from each reaction in a scintillation counter, although a bit tedious, has the advantage of giving a precise figure for each reaction. It should be possible to determine the status of the samples in other less precise ways, such as dot-blotting the released primer onto a membrane such as Whatman DE81 paper and detecting the counts with an imager or on X-ray film. These alternative methods may, however, require the use of a more strongly emitting isotope than ^3H.

5.1.3 Detection of mosaicism

The mini-sequencing technique is sufficiently sensitive that alleles occurring at frequencies of well below 50 per cent can be detected. This is especially useful in the investigation of some forms of cancer, where specific mutations are commonly found in certain tumours (e.g. the K-*ras* codon 12 mutations which are found in a high proportion of colorectal cancers). Such mutations can be used to both diagnose cancer and assess the success of tumour elimination following medical intervention. The solid-phase mini-sequencing technique has been shown to be capable of detecting mutant cells in an environment of normal cells down to a level of 0.05 per cent (Palotie and Syvänen 1992).

6. 5′ nuclease assay

The 5′ nuclease assay is a technique that monitors the extent of amplification in a PCR reaction on the basis of the degree of

◇ Advantages:
● Single step process
● Minimizes contamination risks

◇ Disadvantages:
● Very expensive
● May not detect all mutations

fluorescence of the reaction mix. Low fluorescence indicates no or very poor amplification and high fluorescence indicates good amplification. This system, more commonly used for PCR quantitation, can be adapted to the identification of known mutations, without the need for any post-PCR analysis other than fluorescence emission analysis.

The 5′ nuclease assay is a relatively new method, and as such is not as well tried and tested as many of the other methods. However, it promises to become the method of choice for the detection of known mutations, provided the cost can be reduced.

6.1 The principles

The 5′ to 3′ exonuclease activity of *Taq* polymerase has been utilized to devise a novel system for assaying for PCR amplification. The enzyme cleaves 5′ terminal nucleotides of double stranded DNA. Its preferred substrate is a partially double-stranded molecule, cleaving the strand with the closest free 5′ end. In the 5′ nuclease assay, an oligonucleotide 'probe' which is phosphorylated at its 3′ end so that it cannot act as a DNA synthesis primer is included in the PCR reaction. The probe is designed to anneal to a position between the two amplification primers. When an actively extending *Taq* polymerase molecule reaches the probe molecule, it partially displaces it and then cleaves the probe at or near the single stranded/double stranded junction. The polymerase continues this process of displacement and cleavage until the entire probe is broken up and removed from the template. DNA synthesis is then free to continue (*Figure 4.13*). Perhaps surprisingly, the process does not significantly inhibit amplification (Holland *et al.* 1991).

The imaginative part of the assay is the labelling system which monitors the removal of the probe. The probe is labelled with two different fluorescent labels at different positions (Livak *et al.* 1995a). One label (the quencher) has a localized quenching effect on the fluorescence of the other label (the reporter). This effect is mediated by energy transfer from one dye to the other, but requires the two dyes to be in close proximity. The cleavage of the probe between the reporter and quencher dyes physically separates the two dyes and so results in an increase in fluorescence which is proportional to the yield of the PCR product. Furthermore, the system only monitors amplification of the specific product.

The system is patented and marketed by Perkin Elmer under the trade mark of 'TaqMan'. The probes are supplied by order and use TAMRA (6-carboxytetramethylrhodamine) as the 3′ quencher dye and FAM (6-carboxyl fluorescein), TET (tetrachloro-6-carboxyfluorescein), or HEX (hexachloro-6-carboxyfluorescein) as the reporter dye. The availability of alternative reporter dyes permits multiplex analysis and thus the ability to include internal controls.

The major impact of the assay will probably be in the area of quantitative PCR, i.e. determining the number of target molecules at the

Polymerization

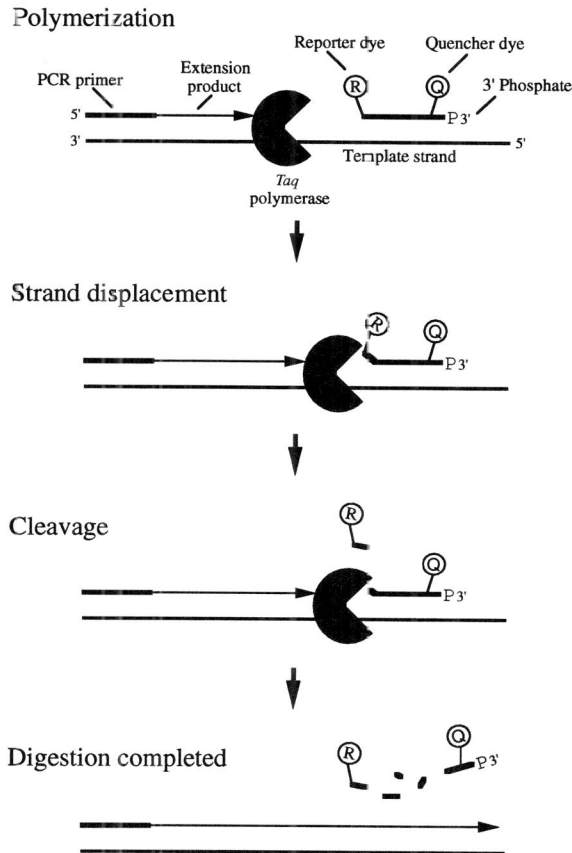

Strand displacement

Cleavage

Digestion completed

Fig 4.13

The 5′ nuclease assay. Displacement and digestion of the probe results in the physical separation of the reporter and quencher dyes, permitting the fluorescence of the reporter dye.

start of the reaction. This function will also be of use in the determination of hemizygosity (allele loss) in human disease processes.

When used for mutation detection, the 5′ nuclease assay does not give cleanly defined positive or negative results, but gives intermediate figures which when analysed permit reliable genotyping. This is done by dividing the reporter fluorescence figure by the quencher fluorescence figure, giving a ratio known as the RQ ratio. Because the emission of the quencher changes little in the PCR amplification, this ratio normalizes for any pipetting errors or volume changes. The RQ ratio of the 'no DNA' control is then subtracted from the sample RQ ratio to give the ΔRQ figure, which represents the amplification of the specific product in the PCR.

6.2   Strategies for mutation detection

There are many different ways in which the 5′ nuclease assay could be used in mutation diagnostics. It is probably adaptable to some of the existing methods, e.g. allele-specific PCR and the oligonucleotide ligation assay. The approaches would be similar, except in allele-specific PCR, amplification would be scored on the basis of fluorescence rather than by gel electrophoresis, and in the oligonucleotide ligation assay, ligation would be scored by a quenching of fluorescence.

However at the time of writing, there have only been two experimental approaches used (Lee *et al.* 1993; Livak *et al.* 1995*b*) and both rely on the direct detection of the mutation by the probe; a mismatch preventing separation of the quencher and reporter dyes.

Method 1. In this approach, the probe (wild-type-specific or mutant-specific) is allowed to anneal equally to both alleles. The quencher dye is positioned not at the 3′ end of the probe, but internally, just 3′ to the mutation position. The two alleles are discriminated on the basis of cleavage separating the reporter and quencher dyes occurring on the fully matched probe, but not on the mismatched probe.

Displacement and cleavage of the fully matched probe occurs in the normal manner. In the case of the mismatched probe, a larger proportion of the probe is displaced before cleavage, such that the cleaved fragment carries both the reporter and quencher, so that the quenching effect is maintained.

Method 2. When the quencher dye is positioned at the 3′ end of the probe it is still possible to differentiate alleles. This approach works surprisingly well, and how it does so is not entirely clear. Even under the conditions needed for the PCR, the probe appears to hybridize more strongly to the fully homologous target, providing some degree of discrimination in itself. Additionally, the mismatched probe molecules are displaced in their entirety without cleavage significantly more easily than the fully matched probe molecules. Together these factors allow sufficient allelic discrimination to permit the genotyping of single base variants.

As ever, in this type of experiment, it is essential to include reaction samples with 'no DNA', samples of known genotype, and to use both wild-type-specific and mutant-specific probes. Instead of conducting separate and complementary reactions for the two probes, the two can be multiplexed in a single reaction, with each labelled with a different reporter dye.

Fully matched probe   Mismatched probe

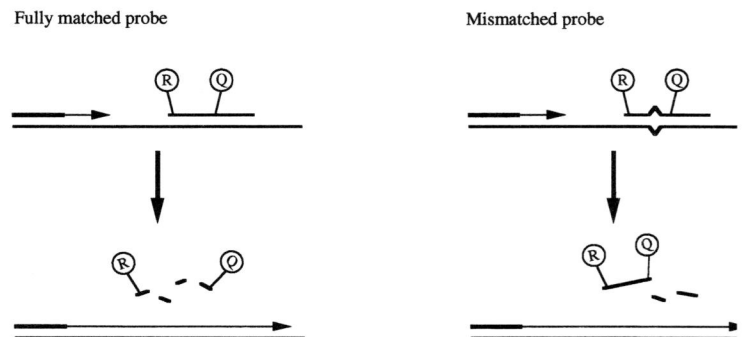

Fig 4.14

The destabilization provided by a single base mismatch prevents cleavage occurring between the reporter and quencher dyes.

One of the most attractive features of the 5′ nuclease assay, is its suitability for automation. When the amplifications are conducted in 96-well microtitre plates, the reactions can be set up by robot. After amplification, the fluorescence is detected by a microtitre-format fluorescence plate reader and automatically analysed by computer. If the microtitre plates are sealed over the top, the amplification products cannot become a contamination threat as there is no need to access the reaction wells. As the assay is a new method, it may take some time before a single globally-used reliable mutation diagnosis procedure is devised.

6.3 Experimental design considerations

In many uses of the 5′ nuclease assay, the quencher molecule is situated at the 3′ end of the primer as this maximizes the probability of cleavage between the reporter and quencher, and is also the position giving maximal hybridization efficiency. The TAMRA (quencher) dye is usually attached via a linker arm fixed to the 4′ position of a thymidine base. This 3′ terminal thymidine base need not be complementary to the target sequence, but complementarity will increase the melting temperature (T_m) (see Chapter 5, Section 4.1). The T_m of the probe should (normally) be at least 5 °C higher than the PCR primers as it is important that the probe is bound to the target when DNA synthesis is occurring. As a general rule, the probe should be 20–30 bases in length with a GC-content of about 50 per cent. Mutations lying in GC-rich DNA may be very difficult to resolve.

The extension temperature in the PCR will usually need to be less than 72 °C as the probe must remain attached during the extension phase. It is therefore usual to perform a two temperature PCR with a combined annealing-extension step. The use of high $MgCl_2$ concentrations (3.5–6 mM) stabilizes probe binding and hence permits higher annealing-extension temperatures, increasing the efficiency of amplification. As with ordinary PCRs, the optimum annealing temperature has to be determined by trial and error.

It is advisable to design the probe such that the terminal 5′ nucleotide is not a guanine base, as this has been shown in some instances to quench the reporter dye fluorescence, even after probe cleavage. As with the design of PCR primers, it is important in the design of the probe to avoid hairpins, self-annealing, primer-probe annealing, runs of single bases, etc. Unlike PCR primers which can be positioned to avoid such problems, the positioning and sequence composition of the probe is restricted. Hence it may not always be possible to use the cleavage prevention by mismatch approach in mutation diagnosis. 'TaqMan' probes in initial experiments for mutation diagnostics did not always work and so probes were designed to some extent by trial and error. Computer programs are now available to facilitate probe design, greatly increasing probe reliability.

7. Comparison of diagnostic methods

Your decision in chosing one of the techniques described in this chapter will probably be based on factors such as logistics, cost, or simplicity. If at all possible, it is always desirable to chose a technique which someone you know has experience of, as even the simplest methods can produce hard-to-solve problems.

If you are working to a very limited budget, you will probably be content to perform the cheapest method despite it perhaps having some disadvantages. The cheapest methods are allele-specific oligonucleotide hybridization (ASO), allele-specific PCR (AS-PCR), and primer-induced restriction analysis (PIRA). ASO as a non-gel based system requires a labelling system of some kind, but has the advantage of being better suited to the analysis of large numbers of samples. AS-PCR and PIRA both require electrophoresis, but this is not difficult and does not require much in the way of expensive equipment. The luxury of gel-free analysis comes at a cost.

For reasons of safety or the lack of a radioactivity usage licence, you may be wanting to avoid using isotopes. AS-PCR, PIRA, and the 5′ nuclease assay do not require isotopes, but ASO, OLA and Mini-sequencing all require a label of some kind. Mini-sequencing, in the form described here, requires only ^{3}H, which is perhaps the safest of all the isotopes used in molecular biology. This technique however can be modified to nonisotopic detection systems. Again, ASO need not be radioactive, but is simpler when radioactive. It is unlikely that you would bother with the radioactive version of the OLA.

If you are inexperienced in the techniques of molecular biology, you may want to choose a technique which is simple to perform. The simplest mutation diagnostic techniques are AS-PCR and PIRA. It remains to be seen just how little experience the 5′ nuclease assay requires. The most tricky of these techniques is probably the OLA. The ligase chain reaction version of this technique is certainly the most difficult approach.

In every molecular biology laboratory, PCR contamination is a serious concern. Extreme care should be taken to minimize this risk, e.g. conducting all pre-PCR and all post-PCR work in different places and using different equipment for the pre- and post-PCR work. Of all the techniques described in this chapter, the 5′ nuclease assay has the greatest to offer in terms of minimized contamination risk as the reaction can be scored following the PCR without needing to unseal the microtitre plate.

In choosing one of these techniques, one of the major factors influencing your decision will be the number of samples you have to process. If the number is small, you probably won't mind using isotopes or electrophoresis. If the number is large, you will be wanting an isotope-free method that is amenable to automation or one which is at least manageable if you have to do it all manually. All of the techniques are suitable to high-throughput analysis to some extent, but ASO, the OLA and the 5′ nuclease assay are the most suitable. The OLA and 5′ nuclease assay are, however, expensive.

All of these techniques are very reliable, but in certain cases particular methods may not work, or only work poorly. A mutation in a palindromic sequence may be a challenge to all of these methods. Regions of very high GC-content may pose a problem to ASO and the 5' nuclease assay. AS-PCR should be feasible in almost all cases, however insertions or deletions in runs of a single nucleotide, cannot be detected. PIRA is not always possible as not every DNA sequence can be manipulated to create or destroy a restriction site in conjunction with a mutation.

Further reading

Allele-specific hybridization

Farr, C. J. (1991). In *Methods in molecular biology, Vol 9: Protocols in human molecular genetics* (ed. C. G. Mathew), pp. 69–76. Humana Press, Clifton, NJ. A good source of experimental detail, including TMAC hybridization.

Handelin, B. and Shuber, A. P. (1995). In *Current protocols in human genetics* (ed. N. C. Dracopoli, J. L. Haines, B. R. Korf, D. T. Moir, C. C. Morton, C. E. Seidman, J. G. Seidman, and D. R. Smith), pp. 9.4.1–9.4.8, Wiley, New York. Provides background, protocols, and trouble-shooting guide.

Allele-specific PCR

Little, S. (1994). *In Current protocols in human genetics* (ed. N. C. Dracopoli, J. L. Haines, B. R. Korf, D. T. Moir, C. C. Morton, C. E. Seidman, J. G. Seidman, and D. R. Smith), pp. 9.8.1–9.8.12, Wiley, New York. A good guide to single and multiplex allele-specific PCR.

Oligonucleotide ligation assay

Landegren, U. (1993). Ligation-based DNA diagnostics. *BioEssays,* **15**, 761. Provides a thorough overview of ligation technology.

Mini-sequencing

Syvänen A.-C. (1994). Detection of point mutations in human genes by the solid-phase minisequencing method. *Clinica Chimica Acta,* **226**, 225. A thorough review of the background, the methodology, and data interpretation.

References

Barany, F. (1991). Genetic disease detection and DNA amplification using a cloned thermostable ligase. *Proceedings of the National Academy of Sciences, USA,* **88**, 189.

Bugawan, T. L., Begovich, A. B., and Erlich, H. A. (1990). Rapid HLA-DBP typing using enzymatically amplified DNA and nonradioactive sequence-specific oligonucleotide probes. *Immunogenetics,* **32**, 231.

Ferrie, R. M., Schwarz, M. J., Robertson, N. H., Vaudin, S., Super, M., Malone, G., et al. (1992). Development, multiplexing, and application of ARMS tests for common mutations in the CFTR gene. *American Journal of Human Genetics,* **51**: 251.

Haliassos, A., Chomel, J. C., Kruh, J., Kaplan, J. C., and Kitzis, A. (1989). Detection of minority point mutations by modified PCR technique: a new approach for a sensitive diagnosis of tumor-progression markers. *Nucleic Acids Research,* **17**, 8093.

Harju, L., Weber, T., Alexandrova, L., Lukin, M., Ranki, M., and Jalanko, A. (1993). Colorimetric solid-phase minisequencing assay illustrated by detection of alpha(1)-antitrypsin z-mutation. *Clinical Chemistry,* **39**, 2282.

Holland, P. M., Abramson, R. D., Watson, R., and Gelfand, D. H. (1991). Detection of specific polymerase chain reaction product by utilizing the 5′ to 3′ exonuclease activity of *Thermus Aquaticus. Proceedings of the National Academy of Sciences, USA,* **88**, 7276.

Ihalainen, J., Siitari, H., Laine, S., Syvanen, A.-C., and Palotie, A. (1994). Towards automatic detection of point mutations—use of scintillating microplates in solid-phase minisequencing. *Biotechniques,* **16**, 938.

Jacobson, D. R. (1992). A specific test for transthyretin 122 (Val–Ile), based on PCR-primer-introduced restriction analysis (PCR-PIRA): Confirmation of the gene frequency in blacks. *American Journal of Human Genetics,* **50**, 195.

Jacobson, D. R. and Moskovits, T. (1991). Rapid, non-radioactive screening for activating ras oncogene mutations using PCR-primer introduced restriction analysis (PCR-PIRA). *PCR Methods and Applications,* **1**, 146.

Jalanko, A., Kere, J., Savilahti, E., Schwartz, M., Syvänen, A. C., Ranki, M., et al. (1992). Screening for defined cystic-fibrosis mutations by solid-phase mini-sequencing. *Clinical Chemistry,* **38**, 39.

Kwok, S., Kellogg, D. E., McKinney, N., Spasic, D., Goda, L., Levenson, C., et al. (1990). Effects of primer-template mismatches on the polymerase chain reaction: Human immunodeficiency virus type 1 model studies. *Nucleic Acids Research,* **18**, 999.

Landegren, U., Kaiser, R., Sanders, J., and Hood, L. (1988). A ligase-mediated gene detection technique. *Science,* **241**, 1077.

Lee, L. G., Connell, C. R., and Bloch, W. (1993). Allelic discrimination by nick-translation PCR with fluorogenic probes. *Nucleic Acids Research,* **21**, 3761.

Little, S. (1994). Amplification-refractory mutation system (ARMS) analysis of point mutations. In *Current protocols in human genetics* (ed. N. C. Dracopoli, J. L. Haines, B. R. Korf, D. T. Moir, C. C. Morton, C. E. Seidman, J. G. Seidman, and D. R. Smith), pp. 9.8.1–9.8.12, Wiley, New York.

Livak, K. J., Flood, S. J. A., Marmaro, J., Guisti, W., and Deetz, K. (1995*a*). Oligonucleotides with fluorescent dyes at opposite ends provide a quenched probe system useful for detecting PCR product and nucleic acid hybridization. *PCR Methods and Applications,* **4**, 357.

Livak, K. J., Marmaro, J., and Todd, J. A. (1995*b*). Towards fully automated genome-wide polymorphism screening. *Nature Genetics,* **9**, 341.

Newton, C. R., Graham, A., Heptinstall, L. E., Powell, S. J., Summers, C., Kalsheker, N., et al. (1989). Analysis of any point mutation in DNA. The amplification refractory mutation system (ARMS). *Nucleic Acids Research,* **17**, 2503.

Nickerson, D. A., Kiaser, R., Lappin, S., Strewart, J., Hood, L., and Landegren, U. (1990). Automated DNA diagnostics using an ELISA-based oligonucleotide ligation assay. *Proceedings of the National Academy of Sciences, USA,* **87**, 8923.

Palotie, A. and Syvänen, A.-C. (1992). Development of molecular genetic methods for monitoring myeloid malignancies. *Scandinavian Journal of Laboratory Investigation,* **53** (S213), 29.

Saiki, R. K., Bugawan, T. L., Horn, G. T., Mullis, K. B. and Erlich, H. A. (1986). Analysis of enzymatically amplified β-globin and HLA-DQα DNA with allele-specific oligonucleotide probes. *Nature,* **324**, 163.

Saiki, R. K., Walsh, P. S., Levenson, C. H. and Erlich, H. A. (1989). Genetic analysis of amplified DNA with immobilized sequence-specific oligonucleotide probes. *Proceedings of the National Academy of Sciences, USA,* **86**, 6230.

Samiotaki, M., Kwiatkowski, M., Parik, J., and Landegren, U. (1994). Dual-color detection of DNA sequence variants by ligase-mediated analysis. *Genomics*, **20**, 238.

Senadheera, D. K., Sapru, M., Williams, J., and Wong, L.-J. C. (1995). Simultaneous screening of more than ten point mutations causing mitochondrial diseases. *American Journal of Human Genetics*, **57**, A1314.

Syvänen, A.-C., Aalto-Setälä, K., Harju, L., Kontula, K., and Söderlund, H. (1990). A primer-guided nucleotide incorporation assay in the genotyping of apolipoprotein E. *Genomics*, **8**, 684.

Syvänen, A.-C., Ikonen, E., Manninen, T., Bengtström, M., Söderlund, H., Aula, P., *et al.* (1992). Convenient and quantitative determination of the frequency of a mutant allele using solid-phase minisequencing: application to aspartylglucosaminuria in Finland. *Genomics*, **12**, 590.

Wall, J., Cai S., and Chehab, F. F. (1995). A 31-mutation assay for cystic fibrosis testing in the clinical molecular diagnostics laboratory. *Human Mutation*, **5**, 333.

Wallace, R. E., Shaffer, J., Murphy, R. F., Bonner, J., Hirose, T., and Itakura, K. (1979). Hybridization of synthetic oligodeoxyribonucleotides to ϕX174 DNA: The effect of single base pair mismatch. *Nucleic Acids Research*, **6**, 3543.

Wu, D. Y., Ugozzoli, L., Pal, B. K. and Wallace, R. B. (1989). Allele-specific enzymatic amplification of β-globin genomic DNA for diagnosis of sickle cell anemia. *Proceedings of the National Academy of Sciences, USA*, **86**, 2757.

5   Scanning mutation detection methods

Detecting an unknown mutation may be likened to finding a needle in a haystack. The ability to find a mutation, which may alter as little as one billionth of the genome, clearly requires a very sensitive technique. Nowadays, that sensitivity is largely provided by the polymerase chain reaction. Sadly there is no single ideal mutation scanning method. As with the mutation diagnostic methods (Chapter 4) each scanning method has its own advantages and disadvantages. For example, one method may detect all mutations but be difficult to perform, whereas another method may miss some mutations but be easy to perform. In this chapter we will cover the theory and methodology behind seven of the most commonly used techniques.

Four of the techniques: RNase cleavage (Section 2), chemical cleavage of mismatch (Section 3), denaturing gradient gel electrophoresis (Section 4), and heteroduplex analysis (Section 5); rely on the formation of a DNA heteroduplex, in which one strand of the DNA is derived from the wild-type and the other from the mutant. The single strand conformation polymorphism assay (Section 1) detects sequence differences in the tertiary structure of single-stranded DNA. The protein truncation test (Section 6) specifically detects nonsense mutations by scoring the size of the protein produced following *in-vitro* transcription and translation. Finally there is the 'sledge hammer' approach of DNA sequencing (Section 7), in which every base pair is individually characterized. At the end of the chapter, the pros and cons of each of these techniques will be compared.

1.   Single-strand conformation polymorphism

◇ Advantages
● Simple

◇ Disadvantages
● 70–98 per cent detection rate
● Detectability cannot be predicted

The detection of point mutations by the single strand conformation polymorphism assay (SSCP) is believed to be due to an alteration in the structure of single-stranded DNA. Mutant single-stranded DNAs are identified by an abnormal mobility on polyacrylamide gels. All types of point mutation can be detected and with apparently equal efficiency. The sequence of the DNA flanking the mutant base does however influence the detectability. The detectability and the banding pattern on the gel cannot be predicted. The overall detection rate is greater than 70 per cent, but if pushed, the rate can excede 95 per cent.

SSCP should not be used as the sole detection method if you need to detect all mutations, or need to declare a piece of DNA mutation-free. The major attraction of SSCP is its simplicity.

1.1 Single-strand DNA conformation

When single-stranded DNA or double stranded DNA is electro-phoresed through a gel matrix, the mobility of the DNA fragment is dependent on its size. Small molecules pass through the pores in the matrix more easily than large molecules and so migrate faster. Conventionally, electrophoresis of single-stranded DNA involves a 'denaturing' gel, which maintains the 'single-strandedness' of the molecules. The denaturant in polyacrylamide gels is typically urea and in agarose gels is typically formaldehyde or sodium hydroxide. The SSCP gel is unconventional in that single-stranded DNA is loaded onto the gel, but the gel does not contain a denaturant i.e. the gel is 'nondenaturing' (Orita *et al.* 1989*a,b*). Running single-stranded DNA on this type of gel permits intramolecular interactions to occur. In other words, the single-stranded DNA is able (partially) to bind to itself. As the DNA is not running as a linear molecule on an SSCP gel, the mobility of the DNA is governed by both its size and tertiary structure (conformation). The tertiary structure of a single-stranded DNA fragment is dependent on the sequence of the entire fragment. If a mutation exists in a given fragment, the conformation will usually be altered.

In addition to the tertiary-structured single-stranded DNA fragments on the SSCP gel, a faster migrating band of double-stranded DNA is also present. This is presumably present because the SSCP gel conditions which permit single-stranded DNA tertiary structure to occur, also allow some re-annealing of the single-stranded DNA into double stranded. This band is not irrelevant as some mutations can be detected as heteroduplex mobility shifts (see Section 5). These heteroduplex shifts can be useful in determining whether an apparent SSCP shift is real or just due to a 'dirty' PCR reaction.

◇ A DNA duplex containing a single base mismatch, can have an altered gel mobility as compared with the wild-type homoduplex.

1.1.1 SSCP gel variables

Unfortunately, the conformation of a single-stranded DNA fragment is dependent on several gel condition variables (Glavač and Dean 1993; Sheffield *et al.* 1993; Ravnik-Glavač *et al.* 1994*a;b*). Not only will the mobility of a given fragment vary depending upon the gel environment chosen, but it will also vary between gels run in supposedly identical conditions on different days. Thus a fragment's mobility is not necessarily repeatable between gels. It is repeatable however, in different lanes of a single gel. Hence a series of DNA fragments of the same sequence run on the same gel, will have identical mobilities. Deviation from this mobility signifies a mutation. The sensitivity of

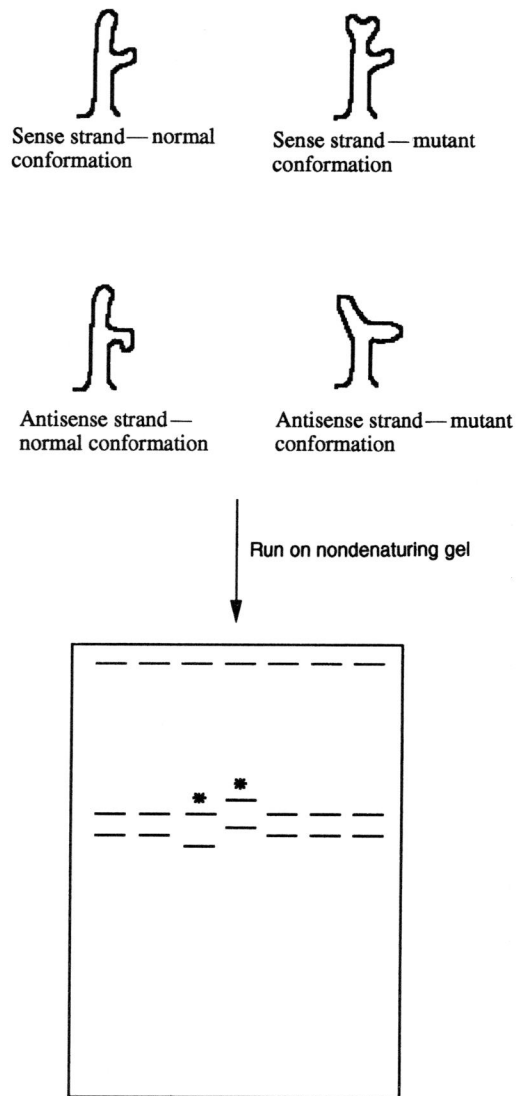

Sense strand—normal conformation

Sense strand—mutant conformation

Antisense strand—normal conformation

Antisense strand—mutant conformation

Run on nondenaturing gel

Fig 5.1

The SSCP assay. Single-stranded wild-type and mutant DNAs form different conformations. When run on a polyacrylamide gel, the mutant samples (marked with asterisks) have an altered mobility.

mutation detection by SSCP depends largely on the gel conditions. These conditions include:

- gel temperature
- cross linker concentration
- acrylamide concentration
- presence of glycerol
- type of gel matrix

Gel temperature

The gel temperature is critical. A high gel temperature will have a denaturing effect, reducing the level of tertiary structure, thus reducing the mutation detection rate. The temperature of the gel can be controlled by both the ambient temperature and the Wattage at which the gel is run. As different mobilities result when gels are run at 4 °C and at room temperature, some SSCP users run gels at both temperatures to maximize their detection rate. There is however, little evidence suggesting that a mutation detected at room temperature might not be detected at 4 °C. It is therefore advisable, if possible, to run the gels in a 4 °C room. If such a room is not available, the room should either be air-conditioned or be a coolish room that doesn't suffer temperature fluctuations. If you are going to be running the gel in a nonrefrigerated room, it is helpful to employ some mechanism for keeping the temperature of the gel low This can be achieved with a fan, in which case a sheet of aluminium should be taped to the surface of the outer glass plate. An alternative method is to run the gel in an apparatus which incorporates a water-cooling system. Keeping the power of the run between 25–40 Watts, should also prevent the gel from getting too hot. The limitation of the Wattage means that the gels cannot be run quickly. It should be possible to run a gel in a 4 °C room in the higher end of the Wattage range, and this will therefore run more quickly than a gel running in a 20 °C room which will have to be run at the lower end of the Wattage range.

Cross-linker concentration

The polyacrylamide gel matrix is formed by the polymerization of acrylamide and a co-monomer cross-linker. The cross-linker used is usually N,N'-methylenebisacrylamide (bis). The pore size of the matrix is governed by both the concentration of the gel (%T) and the proportion of bis in the matrix (%C). For a thorough explanation of gel matrices, see *Electrophoresis: The basics* (Hawcroft 1996).

Conventionally, protein gels contain 2.6%C (37.5 parts acrylamide:1 part bis) or 3.3%C (29:1) and DNA gels 5%C (19:1). For SSCP gels however, it has been demonstrated that C levels of below 5 per cent give better detection rates. C levels of as low as 2 per cent and even 0 5 per cent are now used. Many SSCP users choose 2.6%C as acrylamide/bis can be bought pre-made at this concentration, whereas the lower concentrations cannot.

Acrylamide concentration (%T)

The sensitivity of SSCP detection is also influenced by the acrylamide concentration. DNA fragments for SSCP usually lie within the range of 100–250 bp (see Section 1.1.2). Electrophoretic size separation of single-stranded DNA fragments of this size would normally involve a 5 or 6 per cent polyacrylamide gel. SSCP gels in the past were of the same gel percentage. However, recent studies have shown that higher acrylamide concentrations have the effect of increasing the SSCP mutation detection rate. Some groups use acrylamide concentrations as high as 10 per cent.

Glycerol

Historically, glycerol (a neutral compound) has often been added to SSCP gels. At concentrations of 5 and 10 per cent, glycerol has been shown to enhance the detection rate of SSCP. As with the variable of temperature, some SSCP users have analysed samples on gels with and without glycerol in the hope of maximizing their mutation detection rate. For many years glycerol has been added to SSCP gels as a matter of course, and no one has understood exactly what role it plays. It has been suggested that the effect of glycerol is to reduce the pH of the buffer (TBE buffer) by chemical reaction with boric acid. It is possible that other buffers such as TAE (Tris/acetate/EDTA, pH 8.1) or TBE at a lowered pH might negate the need for glycerol.

MDE gels

◇ MDE gel solution is an AT Biochem product.

MDE (Mutation Detection Enhancement) is a gel matrix which has advantages over conventional polyacrylamide in resolving subtle mobility differences in DNA. It has been used extensively in mutation detection by heteroduplex analysis (see Section 5). Although thorough tests have not been conducted, it would appear that MDE gel has no obvious disadvantages over polyacrylamide in SSCP analysis. MDE does, however, offer the potential advantage of increased detection of double-stranded DNA mobility shifts in the lower part of the SSCP gel. As some mutations can be missed by SSCP analysis, the possibility of detecting the mutation as a double-stranded DNA mobility shift on the same gel is an attractive one. The use of MDE gel should therefore maximize the mutation detection of SSCP gels.

At the time of writing, MDE has not been extensively compared with optimized polyacrylamide SSCP gels. Therefore, polyacrylamide is the gel matrix discussed here in detail.

1.1.2 Other variables

RNA-SSCP

Some studies have suggested that a higher SSCP mutation detection rate is obtained when RNA rather than DNA is analysed (Sarkar *et al.* 1992). RNA molecules can be transcribed from PCR product DNA templates, if bacteriophage RNA polymerase promoter sequences have been incorporated into the 5′ ends of the PCR primers. Typically, the T7 and SP6 RNA polymerases are used (see Section 2). It has been hypothesized that RNA would give better sensitivity than DNA by having a potentially greater repertoire of tertiary structure.

RNA-SSCP has not been extensively compared with DNA-SSCP. If there truly is an advantage, it is probably a subtle one. This, together with the extra work involved in producing RNA transcripts from all the PCR products involved in the SSCP analysis, renders RNA-SSCP less attractive than DNA-SSCP.

DNA fragment length

Another critical parameter in the SSCP mutation detection rate, is the length of the DNA fragment being analysed. In general, the larger the fragment, the less effective SSCP becomes. There is no sudden cut-off point at which mutations cease to be detected, but there is a gradual loss of sensitivity in fragments over about 250 bp. 300 bp is probably the sensible upper limit of fragment size. The optimum size for an SSCP DNA fragment is about 155 bp.

There are several ways to achieve fragment sizes close to the optimum:

1.   Amplify the gene of interest in approximately 150 bp segments.

This is an expensive option, as many oligonucleotide primers will have to be made. It also demands that you have plenty of sample material (DNA or RNA). Problems with this approach are that it is not always possible to design primers at a predetermined distance from each other. Each primer pair needs to be of 'good' sequence and be compatible. It is also necessary that there is an overlap between each PCR product. This is necessary because a mutation which is covered by a PCR primer will be missed (*Figure 5.2*).

If your lab budget is healthy, you can further increase the SSCP detection rate, by creating a higher degree of overlap between PCR products. As stated earlier, strand conformation, and therefore a mutation's detectability, are dependent on the sequence of the entire

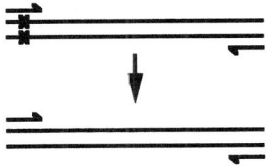

Fig 5.2

A primer lying over a mutation has the effect of converting the mutant base to the wild-type in the amplification.

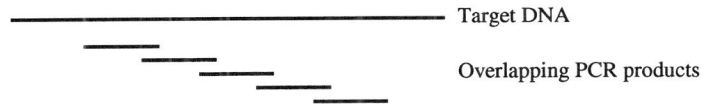

Fig 5.3

Overlapping amplification products.

fragment. If then, there is say a 2× overlap in the PCR products, each base will be examined twice, thus increasing the probability of detection. Apart from the additional cost, this higher overlap approach creates a lot more work for the user.

2.   Amplify the gene in large fragments (1–2 kb) and re-amplify in smaller segments

This option is an alternative to that above, in that it allows for limiting amounts of sample material and, to some extent, permits poor quality PCR amplification.

Modern PCR methodology allows for the amplification of genomic DNA fragments of 2 kb plus. When compared with 100–250 bp products, the amplification of large fragments, is a considerable saving on the amount of sample material used. As such large products are not suitable for SSCP analysis, they need to be broken down into smaller units. One such way is to re-amplify the primary PCR product into overlapping small regions, as described above. The danger of this approach is the increased risk of PCR contamination. If you are considering re-amplification, you should (a) be confident of your PCR cleanliness, and (b) conduct plenty of negative (no DNA) PCR controls.

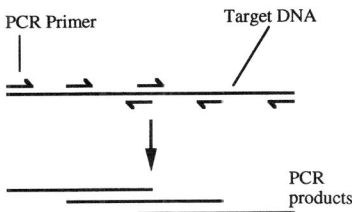

Fig 5.4

A high degree of amplification product overlap.

3.   Amplify the gene in large fragments and digest with restriction endonucleases to produce smaller fragments

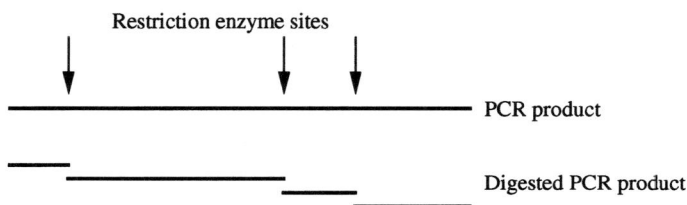

Fig 5.5

Reduction in size of a large PCR product by restriction digestion.

This option is a cheap alternative to that described above. Restriction endonucleases with 4-bp recognition sites, e.g. Sau3AI, RsaI, yield small DNA fragments of an average 256 bp. Digestion with a single enzyme will yield fragments, some of which are smaller than average and some of which are larger. In doing this, you run the risk of your mutation lying in a larger fragment and missing detection. This problem can be ameliorated by digestion with several enzymes. The digestion products can either be run in separate lanes of the gel or combined into a single lane. It is not advisable to end up with a mass of bands in the lane, as some bands may obscure others. You presumably know the sequence of the gene you want to analyse for mutations, and can therefore predict the restriction digest fragment sizes. A rough guideline would be not to exceed more than five restriction fragments per lane. An additional possibility when performing restriction digests for SSCP, is that a mutation may create or destroy a restriction enzyme target site. Should you get this lucky, you will have no problem detecting the mutation!

A recommended strategy

Despite there being so many variables that affect the efficiency of SSCP, a single set of conditions can be used. A compromise has to be made between the amount of work you are prepared to do and what degree of sensitivity you need. The assumption has to be made, that if you are intending to do SSCP analysis, you are willing to forego some sensitivity for the sake of simplicity.

Using the following set of conditions you should be able to detect over 90 per cent of mutations, without too much effort:

● DNA fragments of 150–200 bp
● 8 per cent acrylamide
● 2 per cent C bis cross-linker
● 5 per cent glycerol
● electrophoresis in a 4 °C room

1.2 SSCP methodology

If you have problems with SSCP, it is likely that the problem you have is a PCR problem. The most likely causes of SSCP failure are 'invisible' bands or the presence of extra bands. Both are the result of poorly optimized PCR. As with any PCR amplification, you need a good yield of product with minimal background of nonspecific products.

You should include in the reaction $1\mu Ci$ of $[\alpha\text{-}^{33}P]dCTP$ (1000–3000 Ci/mmol; 10 mCi/ml). The isotope ^{33}P has a β-energy significantly lower than ^{32}P, but a sensitivity of detection greater than ^{35}S. In other words ^{33}P has some of the detectability advantages of ^{32}P and some of the user-friendliness of ^{35}S. Though more expensive than ^{32}P or ^{35}S, it is worth the added cost.

Because the samples are radiolabelled, radioactive precautions must be taken. The dispensing of the undiluted isotope and the addition of diluted isotope to the PCR reaction tubes should be performed behind a perspex radioactivity screen. It is also advisable to perform the post-PCR manipulations behind the screen.

Some SSCP users do not radiolabel the DNA, but stain the gel with ethidium bromide or silver stain. This is not recommended, as to visualize the DNA in this manner, much more DNA has to be loaded onto the gel than when radiolabelled. The loading of less DNA provides better gel resolution and therefore increases SSCP detectability. An additional disadvantage of silver staining is that mutant SSCP bands cannot be excised from the gel for re-amplification (see Section 1.3).

Following the PCR, the sample should be pipetted out of the tube from under the oil and transferred to a new tube. An equal volume of loading dye (95 per cent formamide, 10 mM NaOH, 0.05 per cent bromophenol blue, 0.05 per cent xylene cyanol) should be added and the tube stored on ice, or at -20 °C for up to two weeks

If a primary (nonradioactive) long PCR amplification has been performed, remove 1 μl and add to 100 μl of water. This diluted product (1 μl) should then be re-amplified as above.

◇ Formamide is a denaturant. Bromophenol blue migrates as a dark-blue dye and xylene cyanol as a light-blue dye.

1.2.1 The SSCP gel

The gel apparatus used by most SSCP users is the same as that used for (manual) DNA sequencing. The gel is large and very thin. The main problems are in pouring the gel and in transferring the gel from the apparatus and onto paper. The gel is essentially the same as a sequencing gel, but without urea, i.e. nondenaturing.

To prepare a gel for SSCP, you will require the following reagents:

- acrylamide
- N,N'-methylenebisacrylamide (bis-acrylamide)
- ammonium persulphate
- N,N,N',N'-tetramethylenediamine (TEMED)
- glycerol
- electrophoresis buffer

The gel is formed by polymerization of acrylamide monomers into long chains crosslinked by bis-acrylamide. The resolution of the gel, i.e. the size range of molecules which can be separated, is determined by the concentration of acrylamide monomers in the polymerizing solution.

Acrylamide and bis-acrylamide are both neurotoxins, which can enter the body by inhalation or through the skin. In powder form, they are very light, so easily form an airborne dust when being weighed out. You **must not** inhale such dust. The powder should therefore be weighed out and dissolved in a fume hood or alternatively, bought as a pre-made solution. Dissolved acrylamide, though less immediately hazardous than the powder form, should still be handled with care. If you have made your own acrylamide solution, it is advisable to de-ionize the solution by adding 5 g de-ionizing resin per 100 ml acrylamide solution and gently stirring for one hour at room temperature.

◇ Amberlite MB-1 (BDH) mixed bed ion-exchange resin is recommended for deionizing.

The resin is then removed by filtering the solution. Acrylamide solution is stored at 4 °C and should be good for between three months and one year.

The gel is polymerized by the addition of ammonium persulphate and TEMED to the gel mix. Ammonium persulphate should be made fresh as a 25 per cent solution. TEMED is bought as a liquid and stored at 4 °C. Glycerol should be kept as a 50 per cent stock at room temperature. The final component of the gel mix is the electrophoresis buffer. This provides electrical conductance and has a high buffering capacity.

Preparing the gel assembly

There are many commercially available sets of apparatus for casting and running sequencing sized polyacrylamide gels. Typically, the gels are approximately 40 cm long and 20 cm wide. Each type of gel apparatus will have its own intricacies, but some general rules apply. Possible problems include:

- gel sets before or during pouring
- bubbles form in gel when pouring
- leaks may occur
- the comb may be too thin or too thick

All of these problems can be avoided or cured. The first step in pouring a good gel is to have clean plates. With a wet, nonabrasive paper tissue and a little 'washing-up' type detergent, thoroughly scrub one side of each of the plates. Wash the suds off with water and repeat. The suds should then be washed off again. Wipe the plates with a damp paper tissue—if suds reappear, re-rinse the plates. Dry the plates with dry tissues. The spacers, which separate the glass plates and lie at the sides of the assembly, should be cleaned in the same manner.

◇ The spacers determine the thickness of the gel. For SSCP they should be 0.25 or 0.4 mm thick.

It is advisable to silanize **one** of the glass plates. This provides a very smooth coating to the glass and will (hopefully) ensure that the gel, in its entirety, sticks to the other plate when the plates are separated after electrophoresis. In a fume hood, approximately 2 ml of silanizing solution should be added to the middle of the plate and spread around the surface of the plate in a polishing motion. The plate should be allowed to sit for 10 minutes to dry, before retrieving it from the fume hood and rinsing it with water. Both plates should be

◇ If you are silanizing a plate which has a buffer chamber bonded onto it, you should not let the silane solution touch the bonding seal as it may be damaging.

given a final wipe with 70 per cent ethanol to remove any statically attracted dust. The plates and spacers should then be assembled, making sure the cleaned sides of the plates are inward-facing.

The sides and bottom of the plates then have to be sealed. The nature of this seal will depend upon the type of gel apparatus used. Whichever system is used, be careful that you have sealed the plates properly, as a leaking gel apparatus will probably necessitate recleaning and recasting the gel.

An alternative to the 'sequencing-type' gel are shorter gels such as that in the BioRad ProteanII system, more usually used for running protein samples. This system has the advantage of having a temperature regulation system. Hence a low gel temperature can be maintained at relatively high Wattages. SSCP samples can also be run on the miniature gels of the Pharmacia PhastSystem. It is unclear though, how good this system is at resolving very subtle band shifts.

Pouring the gel

It is wise to find some uncluttered bench space for pouring the gel and to be in a relaxed frame of mind. After the gel has been poured, it needs to rest at an angle of about 30° while it polymerizes—so you will need to find something, e.g. a plastic pipette tip box, for the gel apparatus to rest on. You will also need a means of pouring the gel mix into the apparatus. Some people prefer a 50-ml syringe and others a 25-ml pipette with manual or electric suction. You will also require a gel comb. Combs seem to vary in width, so it is worth having a few combs handy and using the one which fits best. If using a gel apparatus with a built-in upper buffer chamber, you should either block the top of the chamber with a tissue to stop acrylamide spilling down into it, or half fill the chamber with water so that if acrylamide spills into it, it does not set.

Ammonium persulphate and TEMED can now be added to the gel mix, and the beaker gently swirled to mix. Beware, adding too much ammonium persulphate and TEMED may cause the gel to set before you have finished pouring it. The gel apparatus then needs to be held upright. It is important that you are comfortable in the position, as the gel should take 2–3 minutes to pour. For details on pouring the gel, see *DNA sequencing: The basics* (Brown 1994).

Once the gel has been poured, you need to insert the comb. This is usually a sharkstooth comb (*Figure 5.6*) and it is inserted upside-down. Its purpose being to create a flat surface at the top of the gel. The comb should be pushed down to 3–5 mm below the top of the smaller plate. The gel apparatus should now be left to rest at an angle of about 30 °, until it has set. When any remaining unpoured acrylamide in the beaker has set, it is safe to assume that the gel itself has set. It is prudent though, to wait another 15–30 minutes before beginning to run the gel. Once polymerized, the seal can be removed from the bottom end of the gel.

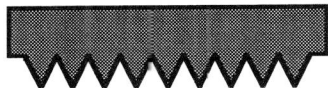

Fig 5.6

A sharkstooth comb.

1.2.2 Loading and running the gel

The gel apparatus is now ready to be attached to the full gel running assembly. If you are going to run the gel in a 4 °C room, you should

◇ 10x TBE = 0.89 M Tris base, 0.89 M boric acid, 20 mM EDTA. As a rule of thumb, add 1 μl ammonium persulphate and 1 μl TEMED per ml acrylamide.

◇ Combs often have holes in them which enable something such as a spatula to lever the comb out, should it be difficult to remove.

◇ Make sure you know the loading order of the samples.

◇ Loss of buffer can result in electrical 'arcing' and is a serious fire hazard.

allow the gel to equilibrate to the temperature of the room, prior to starting the run. Once the complete gel apparatus is assembled, the buffer chambers should be filled with 0.5× TBE. The comb should then be removed, rinsed, and replaced, this time with the sharkstooth side downwards. The teeth should be pushed in only far enough to break the surface of the gel. Be very careful when replacing the comb. The combs are expensive and it is very easy to catch one or more teeth on the lip of the smaller plate, damaging the comb.

Loading your samples and running the gel

Prior to loading, the wells should be flushed with a Pasteur pipette, removing any buffer crystals or glycerol which may have leached out of the gel. The gels are loaded with standard laboratory micropipettors and ordinary 'yellow' (2–200 μl) tips. The pipettor tip is placed immediately above the well and the sample displaced very gently. Gently and carefully withdraw the pipettor, avoiding sucking the sample upwards and out of the well, and also avoiding catching the comb with the tip and dislodging the comb.

It is prudent prior to running your samples that you load 1–2 μl loading dye into three or four wells distributed across the gel. These should be left for 2–5 minutes to check that the samples will not leak sideways between wells. Any leakage will be due to the comb being too loose. If leakage does occur, you can try pushing the comb further into the gel and retesting. Leakage greater than about 10 per cent of the sample is bad news for SSCP gels, as it is preferable to load all the samples in one go. However, the time the samples spend sitting in the gel prior to electrophoresis can be reduced by only loading half the gel and applying current for 5 minutes to get the samples into the gel. Then the remaining samples can be loaded. If serious leakage occurs, the gel will have to be abandoned. The dye used as leakage tester can be removed by flushing with a Pasteur pipette, or by running the dye into the gel.

Now that the gel apparatus is ready, you can begin to prepare your samples for running. The samples should be incubated in a water bath at 80–95 °C for 5 minutes. The samples should then be placed on ice and then loaded in 1–2-μl volumes onto the gel.

The gel should be run at 25–40 W for between 6 and 16 hours. If the run is expected to take 6–8 hours, it may be wise to prepare the gel on the previous day. If you choose to store the gel overnight, you should cover exposed ends of the gel (i.e. the top and possibly the bottom) with wet paper tissues to prevent the gel from drying out. The tissues should then be covered in cling film and the gel should be stored at 4 °C. If the gel is to be run overnight, make sure that neither buffer chamber leaks and that both are full. When all the samples have been loaded, connect the leads to the power pack and the gel assembly, and turn the power on. Make sure the leads are plugged into the correct terminals, i.e. red–positive, black–negative.

The time required for the run depends upon the time taken by the single-stranded DNA to migrate approximately half-way down the gel. Running them much further will cause the bands to become fuzzy.

Running them less than one third of the way down the gel will not provide sufficient resolution. The exact gel running time can only come from experience. For your first run however, running the dark blue (bromophenol blue) dye to the bottom of the gel would be sensible. With experience you will learn where each dye should be at the end of the run.

Dismantling the gel apparatus

When the run is complete, the power pack should be turned off and the leads removed. The gel plate assembly should be removed from the rest of the apparatus. The lower buffer chamber should be discarded down an approved radioactive sink. Should you spill any of this buffer, you must immediately wipe it up, check for radioactive contamination, and dispose of the tissues in the radioactive bin.

In order to get an autoradiographic exposure of the gel, it is necessary to transfer the gel from the plates and onto blotting paper. The separation of the plates is the final stage at which you run the risk of ruining the experiment. For details on how to successfully transfer and dry the gel, see *DNA sequencing: The basics* (Brown 1994).

Autoradiography

You should place the dried gel in an X-ray cassette and take it, together with a box of X-ray film to a dark room for loading. It is safe to use a red safe-light for this purpose. You should not wear gloves, as this can cause static electricity images on the film and you should not crease the film as creases do not flatten out and cause an image on the film. The film should be laid over the gel, and the cassette closed.

The film should be exposed overnight in the first instance. Longer exposures should be performed if necessary.

1.3 The result

An ideal result is illustrated in Figure 5.7. Mutations can be observed as clear SSCP band shifts in lanes 1 and 2 of gel 1 and as subtle band shifts in lanes 1 and 7 of gel 2. In these cases, all mutations can also be observed as double-stranded heteroduplex band shifts. The importance of these bands is emphasized by the heteroduplex shift in gel 2 lane 9, where no SSCP band shift is visible.

Conventionally the mutations are characterized by re-amplifying the mutant samples followed by cloned or direct sequencing. It is also possible to characterize the mutation by re-amplifying the shifted band from the gel. When doing this, you should also attempt to re-amplify a part of the gel which has no band, as a contamination control. Successful amplification of the mutant band eliminates the problems of sequencing a heterozygous mutation (see Section 7, this chapter). An alternative strategy is to clone the PCR product and subject the cloned inserts to SSCP prior to DNA sequencing, to identify the clones carrying the mutant allele.

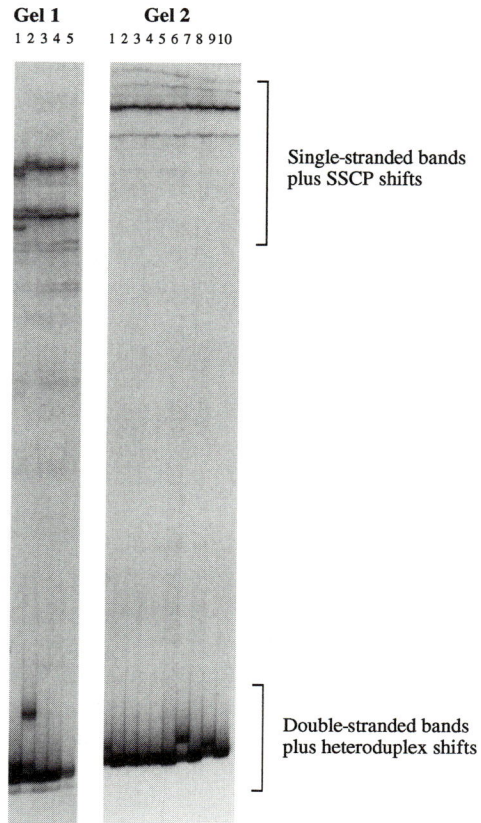

Gel 1   Gel 2

Single-stranded bands
plus SSCP shifts

Double-stranded bands
plus heteroduplex shifts

Fig 5.7

An autoradiograph of two SSCP gels. Gel 1 shows two mutations (lanes 1 and 2), one of which shows a clear heteroduplex shift (lane 2). Gel 2 shows three mutations (lanes 1, 7, and 9). None give obvious SSCP shifts, but are revealed by heteroduplex shifts. The photograph was donated by Dr A. Schafer.

1.3.1 Problems

The most likely problems in SSCP are a blank or very faint autoradiograph or too many bands. Both are due to poor quality PCR amplification. Lack of bands may be cured by replacing the PCR reagents or changing the primers. Faint bands may be enhanced by increasing the amount of isotope used or decreasing the amount of cold dCTP in the reaction.

If additional bands are observed, the PCR needs to be cleaned up. This may be done by altering the primer concentration, enzyme concentration, magnesium concentration, the annealing temperature, or cycle number. It is advisable to make sure you can produce a clean band in a 'cold' PCR before starting the SSCP.

1.4 Application of fluorescence technology to SSCP

SSCP is amenable to fluorescent analysis (Iwahana *et al.* 1994). The main difference between isotopic SSCP and fluorescence SSCP is that in the latter, the PCR products must be end-labelled. Incorporation of fluorescent nucleotides into the PCR product results in smeary bands on the gel. The labelling can be achieved in several ways, including:

1.   The synthesis of dye-labelled primers. This is the simplest, but most expensive option.

2.   By Klenow end-labelling the PCR product. This is the cheapest option, but requires several steps.

3.   By synthesis of primers with M13 tails. This option is cheaper than the synthesis of dye-labelled primers, but requires a secondary PCR amplification. The primary PCR primers are made with 17-bp tails on the 5′ end, so that the PCR product may be re-amplified with dye-labelled M13 forward and reverse (sequencing) primers. In this way, you only need have a single pair of dye-labelled primers for all your needs.

Apart from the elimination of radioactivity, fluorescent SSCP has the advantages of being able to include internal lane standards and accommodate more samples per gel by different colour multiplex loading.

Preliminary SSCP experiments have been performed on the Applied Biosystems and Pharmacia fluorescence-based semi-automated sequencing machines. The ABI 373 machine is poorly suited to SSCP because the temperature of the gels cannot be controlled, necessitating very long, low-Wattage runs. The newer ABI 377 machine, which does have temperature control is much more promising.

2.   RNase cleavage

◇ Advantages
● Scans up to 1 kb
● Maps mutation

◇ Disadvantages
● ~70 per cent detection rate
● Radioactive

Ribonuclease A specifically digests single-stranded RNA. The enzyme can also cleave heteroduplex molecules at the point of a mismatch. The extent of cleavage at single base mismatches is not only dependent on the type of mismatch, but also on the sequence of DNA flanking the mismatch.

Mutations are detected as fragments smaller than the uncleaved heteroduplex on denaturing polyacrylamide gels. The overall point mutation detectability, is about 70 per cent or so.

RNase cleavage has become a rather unfashionable mutation detection technique. The elimination of radioactivity, could however make the technique more fashionable in the future.

2.1   The theory

Various ribonuclease enzymes, including RNase A, RNase T1, and RNase T2 specifically digest single-stranded RNA. When RNA is annealed to form double-stranded RNA or an RNA/DNA duplex, it can no longer be digested with these enzymes. However, when a mismatch is present in the double-stranded molecule, cleavage at the point of mismatch may occur. The most commonly used and studied ribonuclease for mutation detection is RNase A.

The technique is based upon forming a heteroduplex between a radiolabelled single-stranded RNA probe (riboprobe) and a patient-derived PCR product. The resulting heteroduplex is an RNA/DNA hybrid molecule. When treated with RNase A, if a point mutation is present, the RNA strand of the duplex may be cleaved (Myers *et al.* 1985; Winter *et al.* 1985). The sample is then heated to denature and run on a denaturing polyacrylamide gel. If the RNA probe has not been cleaved, its size will be that of the PCR product. If the probe has been cleaved, its size will be smaller (*Figure 5.8*). The technique is essentially identical to 'RNase protection', in which RNA or DNA fragments are detected, quantitated, or sized by annealment to an RNA probe.

As with the other mutation detection techniques, deletions as small as 1 bp are easily detectable. Small insertions may not be as easily detected as small deletions, as 'looping-out' occurs on the target strand rather than the probe strand. The detectability of base substitutions is highly variable. The extent of cleavage is governed by not only the type of mismatch but also the sequence context of the flanking DNA. Thus the detectability of a given point mutation cannot

Fig 5.8

The RNase cleavage methodology.

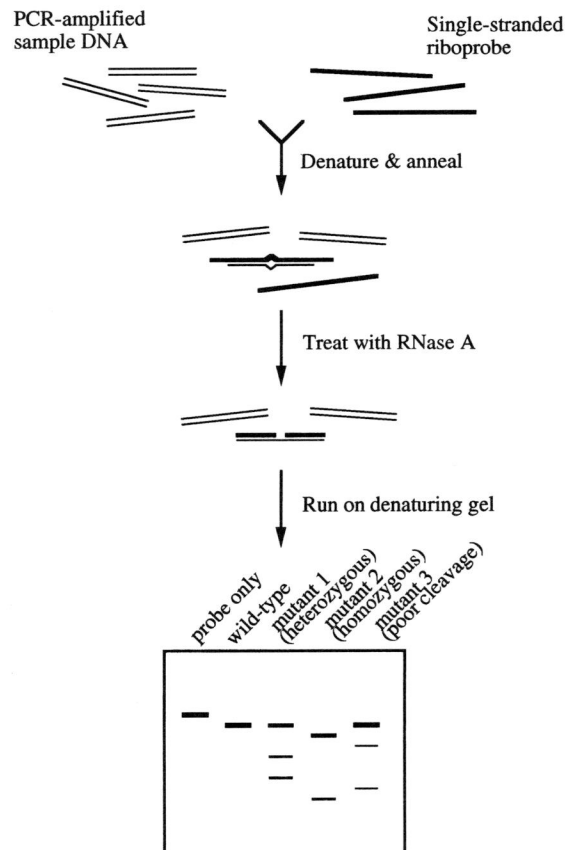

be accurately predicted. When only one strand of the PCR product is analysed, the mutation detection rate is only about 35 per cent. When both strands are analysed, i.e. with sense and antisense riboprobes, the detection rate is doubled to about 70 per cent. Much of the work to define the efficiency of the technique was performed prior to the advent of PCR technology, when the method was used to analyse genomic DNA and cellular RNA. It may therefore be that the mutation detection rate is significantly higher than 70 per cent when large amounts of target DNA, such as PCR products, are analysed.

For the same reasons as in mutation detection by chemical cleavage of mismatch (Section 3), relatively large regions can be analysed (up to 1 kb). Mutations are mapped in the same way as with uniformly labelled probes in CCM analysis, facilitating the characterization of the mutation (Section 3.1.1).

2.1.1 Riboprobe synthesis

Highly efficiently labelled single-stranded riboprobes can be produced from plasmid templates containing the bacteriophage SP5, SP6, T3, and T7 RNA polymerase promoters. Wild-type DNA fragments, derived from either PCR products or library clones, are subcloned into transcription vectors such as pBluescript (Stratagene) or pTZ18/19 (Pharmacia). The recombinant plasmid should be linearized with a restriction endonuclease at the opposite end to which the bacterio-phage promoter to be used lies. This permits the probe to be derived from only the insert and not the insert and the plasmid (*Figure 5.9*).

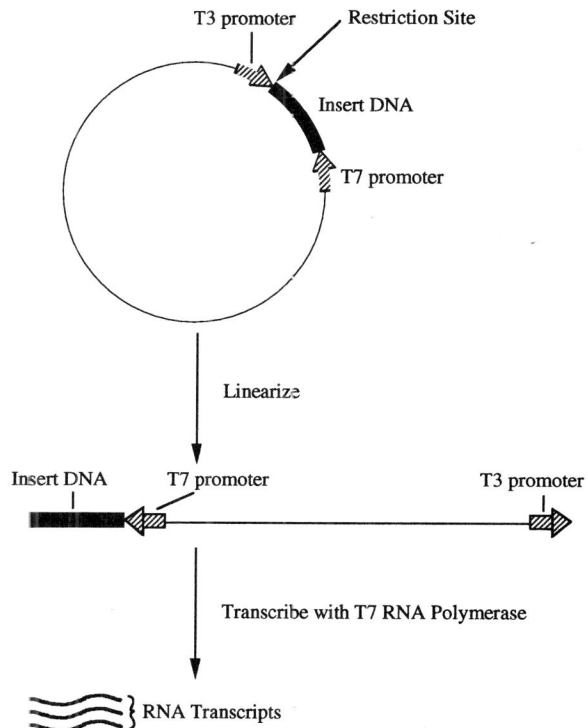

Fig 5.9

In-vitro transcription from a cloned template.

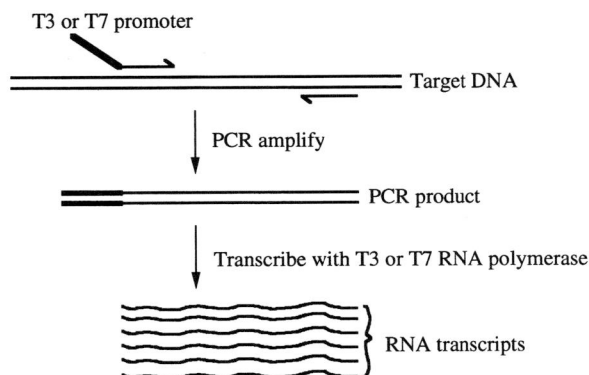

Fig 5.10

In-vitro transcription from a PCR product template.

The T3 or T7 polymerases are used to generate single-stranded RNA transcripts derived from the insert DNA. The transcripts should all, or mostly, be full-length. When one of the NTPs in the reaction mix is labelled, the transcripts generated are 'hot' riboprobes. Transcripts derived from the other strand of the plasmid insert, can be produced by subcloning in the reverse orientation or using the alternative RNA polymerase (T3 polymerase in *Figure 5.9*), following linearization at the other end of the insert.

If the riboprobe is longer than the target PCR product, the overhanging ends will be digested away. Any unhybridized probe will be digested to completion. The wild-type PCR product/riboprobe duplex will not be internally cleaved and is thus 'fully protected'. The mutant PCR product/riboprobe duplex is cleaved at the mismatch to produce smaller bands or 'cleavage products' on the gel. The gel is viewed by autoradiography and therefore only the probe part of the duplex is visualized.

An alternative probe generation method, which avoids subcloning, is to incorporate T3 or T7 promoter sequences into the 5′ ends of PCR primers. In this way, the PCR products contain the relevant promoter sequence at one end, so that the RNA transcripts can be driven directly off the PCR product (*Figure 5.10*). This method is gaining increasing favour, and is now perhaps the most commonly used method of RNA probe generation.

Table 5.1 Efficiencies of mismatch cleavage by RNase A[a]

RNA:DNA	Cleavage efficiency (per cent)
C:A	50–100
C:C	100
C:T	50–100
U:G	0–90
U:C	5
U:T	25–75
G:A	0–75
G:T	0–15
G:G	0
A:C	0–100
A:A	0–20
A:G	0–100

[a] Data is taken from a study by Myers *et al.* (1985).

Probe design

The rules of RNase cleavage probe design also apply to chemical cleavage of mismatch (Section 3, this chapter). Probes should be between 250 and 1200 bp in length. The ideal probe length is about 600 bp. Larger probes tend to give higher levels of background. Probes should also be overlapping to avoid missing mutations at the ends of the fragment.

A disadvantage of the RNase cleavage method is the variability in probe cleavage efficiency (*Table 5.1*). Some mismatches are cleaved better than others, but there is also variability in cleavage efficiency of a particular mismatch. This is due to the effect of context differences. *Figure 5.11* illustrates variable mismatch cleavage efficiencies and the importance of using both sense and antisense probes. A 'T to G'

T to G mutation

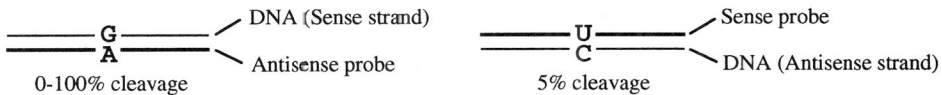

C to G mutation

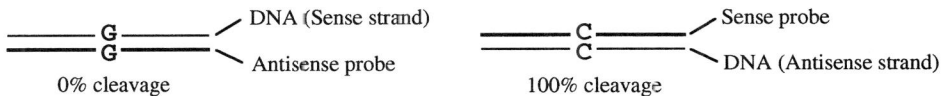

Fig 5.11

The importance of using both sense and antisense probes.

mutation may well go undetected when the sense probe is used (a U:C mismatch). When the antisense probe is used, the mutation is more likely to be detected (an A:G mismatch). A 'C to G' mutation will probably go undetected when the antisense probe is used (a G:G mismatch), but will almost certainly be detected when the sense probe is used (a C:C mismatch). There is little evidence that the RNases T1 and T2 can cleave mismatches resistant to cleavage by RNase A.

2.2   RNase cleavage methodology

2.2.1   Riboprobe synthesis

Recombinant plasmid DNA should be purified by caesium chloride centrifugation or over a commercial cleaning column, e.g. Qiagen. The DNA should be cut to completion with a restriction endonuclease. Linearized DNA should be cleaned by phenol/chloroform extraction, followed by chloroform extraction and then ethanol precipitation. PCR product probe templates should also be cleaned, preferably by passing through a cleaning column.

After labelling (riboprobe synthesis), the probe is digested with RNase-free DNase I to remove the template DNA. Following phenol/chloroform and chloroform extractions, the probe is ethanol precipitated with ammonium acetate as the salt (see Chapter 4, Section 4.2). The probe pellet, which may be invisible, is resuspended in water. The pellet must not be dried before resuspension. Dissolving the pellet may require much pipetting up and down. In initial experiments, the activity of the riboprobe should be quantitated by counting an aliquot in a liquid scintillation counter. In subsequent experiments, measuring the activity of an aliquot (e.g. 1 μl of a 1/1000 dilution, spotted onto paper and dried) with a Geiger counter should be sufficient.

2.2.2   Handling RNA

RNA degrades very easily. To avoid contaminating the RNA with RNases, you need to be meticulously clean. Gloves should always be worn, all glassware and solutions should be treated with diethyl

◇ In producing a riboprobe, much manipulating of the reaction tube is required. In order to minimize radioactive exposure to the fingers during these manipulations, precautionary steps should be taken. Whenever possible, the tube should be held in a perspex block or with forceps. Holding the tube directly with the fingers, subjects them to a large radiation dose. A shield fixed to the barrel of the pipette may also provide some protection to the pipetting hand.

◇ DEPC is irritating to the eyes and skin. Its vapour should not be inhaled. It should therefore be handled with care in a fume hood.

pyrocarbonate (DEPC). DEPC-treatment involves adding 1 drop of DEPC from a Pasteur pipette per 100 ml of solution. The solution should be shaken vigorously and allowed to stand overnight before autoclaving. Tris solutions are not DEPC-treatable. In initial experiments it is advisable to soak pipette tips and Eppendorf tubes in DEPC-water to ensure cleanliness.

2.2.3 Generating the heteroduplex

◇ Formamide is deionized in the same way as acrylamide (see Section 1.2.1).

To generate the probe–PCR product heteroduplex, the two samples are mixed and ethanol precipitated. The pellet is resuspended in good quality deionized formamide. When the pellet is fully dissolved hybridization buffer is added and the sample heated at 90 °C for 10 minutes to denature, and allowed to anneal at 45 °C for 1 hour.

2.2.4 RNase cleavage

The heteroduplexed sample is next treated with RNase A. The activity of the enzyme can vary between batches and so it is worth making a large stock. RNase A master stock is prepared by dissolving at 2 mg/ml in water. The solution should be boiled for 10 minutes to remove contaminating DNases, before storing at –20 °C in aliquots. Working stocks of 5–100 μg/ml are made by diluting the master stock.

Cleavage is conducted by simply adding the RNase A to the heteroduplex, and incubating at room temperature. The optimal level of digestion will have to be determined experimentally. For the initial experiments at least, a range of enzyme concentrations, or a time course of a single concentration is recommended. The use of probes longer than the target PCR product are useful in defining enzyme under-treatment, as the overhanging ends of the probe should be cleaved off.

At high enough concentrations, the enzyme will begin to digest homoduplexed RNA. It is this, and the physical breakdown of the probe, known as radiolysis, which gives the background on RNase cleavage gels. The background is present as a smear on the gel, together with discrete faint bands. These bands do not get confused with cleavage products, as they are identical from sample to sample. It is advisable to always have at least some background. Over-digestion of the duplex will maximize the degree of cleavage at a mismatch. Thus a sufficient amount of enzyme should be used to give some background, but not so much so that the background is too intense to be able to see any cleavage products. Additional background will be present if the sample PCR reactions are not clean. The RNase digestion is stopped by incubation with SDS and proteinase K.

◇ Loading dye is: 0.1 per cent bromophenol blue, 0.1 per cent xylene cyanol, 10 mM EDTA, 1x TBE in deionized formamide.

It is advisable in initial experiments to phenol/chloroform extract and ethanol precipitate the proteolysed sample. To aid precipitation, 1 μg tRNA should be added. The sample is then dissolved in loading dye. Once the RNase cleavage technique is working, you might find that it is possible to omit the precipitation step without it having a detrimental effect.

2.2.5 Preparing the gel

The gels for RNase cleavage analysis are denaturing polyacrylamide gels (19:1 acrylamide:bis) (see Section 3.3.3). It is advisable to include 20 per cent formamide as well as 7 M urea to ensure denaturation is maintained. The acrylamide concentration will depend upon the probe size. The buffer in the gel and the buffer chambers is 1× TBE.

Any polyacrylamide gel apparatus is suitable, however the Bio-Rad ProteanII system is ideal, as warm water can be circulated to help maintain the gel at a constant hot temperature. Gel plates should be cleaned as described in Section 1.2.1 and spacers should be 1 mm thick. If a sequencing gel apparatus is to be used, a well-forming comb rather than a sharkstooth comb should be used.

2.2.6 Running the gel

The samples should be heated at 95 °C for 10 minutes to denature and placed on ice. The samples should then immediately be loaded onto the gel. If the samples were not precipitated, the maximum amount should be loaded. Size markers should be included and the gel should be run as described in Section 3.3.3.

When the run is complete, the gel assembly should be dismantled and the plates separated. The contents of the lower buffer chamber should be poured down a radioactivity disposal sink. If the probe was not precipitated, this might be very 'hot'. You should check to see how 'hot' the gel is with a Geiger counter. If it is more than about 50 counts, the gel should be handled behind a screen. It should not be necessary to 'fix' the gel (see Section 3.3.3). The gel should be covered with Saran Wrap™ and dried onto 3MM paper on a gel drier. The dried gel should be exposed to X-ray film overnight using an intensifying screen at –80 °C.

2.3   The result

An ideal result is shown in *Figure 5.12*. Cleavage is observed in tracks 1 and 6. The bands observed in all tracks are background bands.

2.3.1 Problems

◇ Gel purification of the probe should be avoided if possible, as it is tricky and very radioactive.

Under- and over-digestion with RNase A can cause the missing of mutations and the absence of protected bands, respectively. Failure to generate homogeneous full-length probes will add to the background. An aliquot of the probe alone should always be included on the gel, to determine its integrity. If it proves very difficult to obtain a full length probe, the probe can be purified by running on a gel and eluting it from a gel slice. Failure to generate a very hot probe is likely to be due to an unclean DNA template.

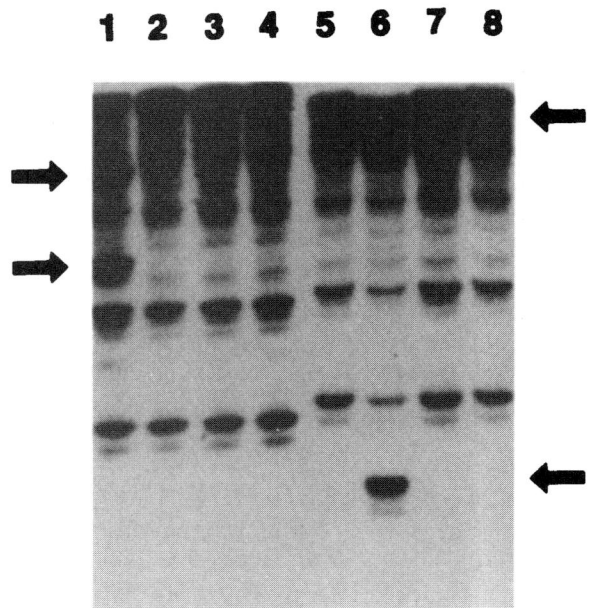

Fig 5.12

An RNase cleavage gel. Arrows indicate cleavage products present in lanes 1 and 6. The bands present in all tracks are background bands. (Reproduced with permission from Miyoshi *et al.* 1992.)

2.4 Nonisotopic RNase cleavage

As RNase cleavage is unfashionable as a mutation detection technique, little effort has been made to modify and improve it. Modification to remove the isotope would make the technique very much more attractive. The use of ^{33}P labelling would make a significant improvement. The US biotechnology company Ambion are currently marketing, in addition to kits for isotopic RNase cleavage analysis, kits for performing unlabelled analysis in which the fragments are visualized by ethidium bromide staining. The gels however appear 'dirtier' than the autoradiographs of isotopic RNase cleavage.

Just as chemical cleavage analysis is being successfully adapted to fluorescence labelling, RNase cleavage could probably also be adapted. Fluorescently labelled ribonucleotides though are not widely available or actively marketed.

3. Chemical cleavage of mismatches

◇ Advantages
● 100 per cent detection
● Scans up to 1 kb
● Maps mutation

◇ Disadvantages
● Complex multistep procedure
● Toxic chemicals

The detection of mutations by chemical cleavage of mismatch (CCM) is the most effective of the mutation scanning techniques. It relies upon the chemicals hydroxylamine and osmium tetroxide to react with the mismatch in a DNA heteroduplex. Subsequent treatment with piperidine cleaves the heteroduplex at the point of mismatch. Mutations are detected as fragments smaller than the untreated heteroduplex on denaturing polyacrylamide gels.

The probable 100 per cent detection rate, coupled with the ability to scan DNA fragments up to 1 kb in size, make CCM seem the ideal mutation detection method. If you need to detect all mutations or declare a piece of DNA mutation-free, CCM does provide an alternative to DNA sequencing. Before deciding on this method you should however check out the Health and Safety information on the chemicals used. These chemicals are nasty. It is also worth noting that, if you do not have access to fluorescence technology, most of the (many) manipulations involved will be radioactive.

To justify a foray into CCM analysis it is therefore likely that you require 100 per cent detection or need to screen a long stretch of DNA.

3.1 The theory of chemical cleavage of mismatch

◇ Both hydroxylamine and osmium tetroxide attack the pyrimidine 5,6 carbon–carbon double bond, permitting cleavage by piperidine.

The chemistry of CCM is an extension of that used in Maxam–Gilbert DNA sequencing, in which piperidine is used to cleave T and C bases following reaction with hydrazine. CCM was developed through an experiment to find chemicals which would react with mismatched

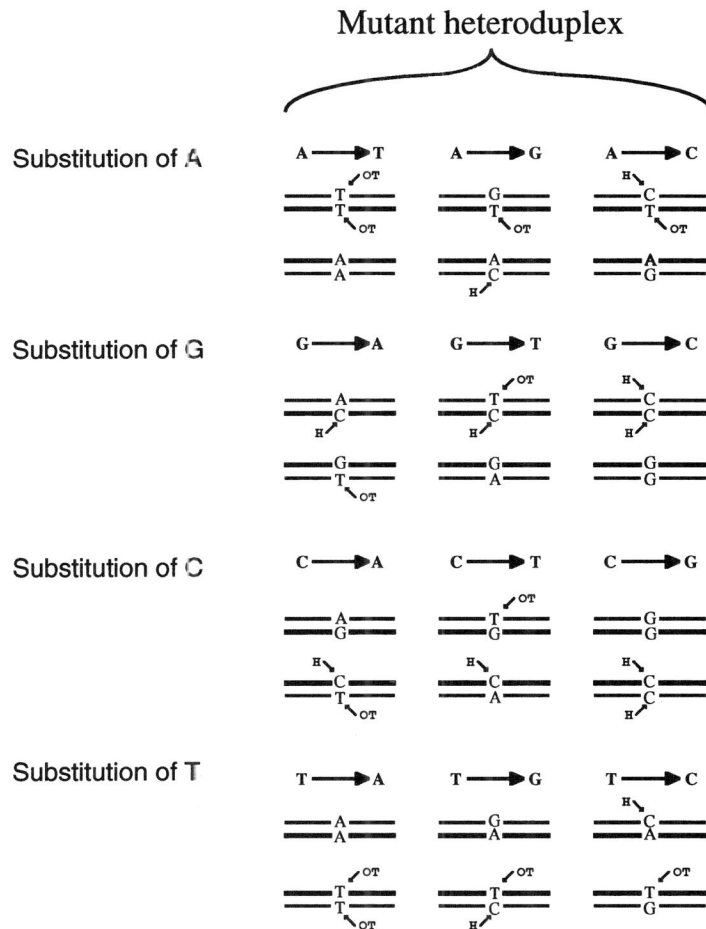

Fig 5.13

Each possible single base pair mismatch is recognized in at least one of the two heteroduplexes. OT = osmium tetroxide; ⊣ = hydroxylamine.

G to T mutation

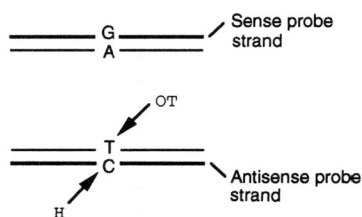

Fig 5.14

In the case of a G to T mutation, the mutation is only detected in one of the two heteroduplexes.

C to T mutation

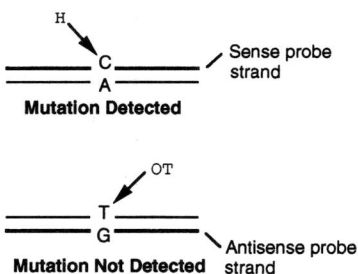

Fig 5.15

When the probe is wild-type DNA, a C to T mutation is only detected in the heteroduplex containing the sense probe.

purine and pyrimidine bases in a heteroduplex, permitting cleavage of the DNA by piperidine (Cotton *et al.* 1988). Hydroxylamine was found to react with mismatched C bases and osmium tetroxide with mismatched T bases. Mismatched G and A bases are detected by virtue of the fact that their complementary C and T bases are reactive. In this way, all types of single base mismatch can be detected.

In the CCM reaction, radiolabelled wild-type DNA (the probe) is mixed with the sample DNA to form heteroduplexes. If both strands of the probe are labelled, then all types of mismatch are detectable. For example, if a G to T mutation occurs on the sense strand, the mutation will not be detected when the probe is the sense strand, but will be detected when the probe is the antisense strand (*Figure 5.14*).

It is important to realize that cleavage is only observed on the probe strand. The target strand is unlabelled and therefore not observed on the autoradiograph of the gel (*Figure 5.15*).

Analysis of a number of mutations detected by CCM has shown some T bases in T–G mismatches to be resistant to osmium tetroxide modification, which are therefore not cleaved. In these cases the sequences flanking the mutation are thought to be important. This partial failure of the technique is countered, however, by the accompanying C–A mismatch, which is reliably modified by hydroxylamine.

Hydroxylamine and osmium tetroxide, in addition to modifying mismatched Cs and Ts respectively, do also modify matched Cs and Ts, albeit at a much lower level. It is therefore essential in the experimentation that a compromise between enough reaction time to generate sufficient modification and not so much time as to generate a high background is achieved.

Matched bases close to the mismatch tend to be more reactive than those further away. This is probably due to loosening of the duplex at the region of mismatch or due to loosening generated by modification of the mismatched base. It has also been shown that when the reactive mismatched base lies in a run of three or more bases of the same type, e.g. CCCC or TTTT, the matched bases show higher than usual reactivity.

As with the other mutation scanning methods, insertions and deletions as small as 1 bp are easily detectable. Deletions are detected by cleavage of an unmatched C or T in the probe strand. The increased reactivity of matched bases nearby will contribute to the cleavage efficiency. Insertions are detected indirectly due to the increased reactivity of flanking matched C and T bases. As with point mutations, the detectability is maximized by the use of probes of both senses. Additional sensitivity to insertion mutations can be achieved by the use of not only wild-type probes but also of mutant probes against wild-type target which effectively converts the insertion into a deletion.

3.1.1 Mutation mapping

Not only is CCM highly efficient in detecting mutations, but being a cleavage method it also maps the mutations detected. The size of the bands (cleavage products) observed on the autoradiograph will indicate

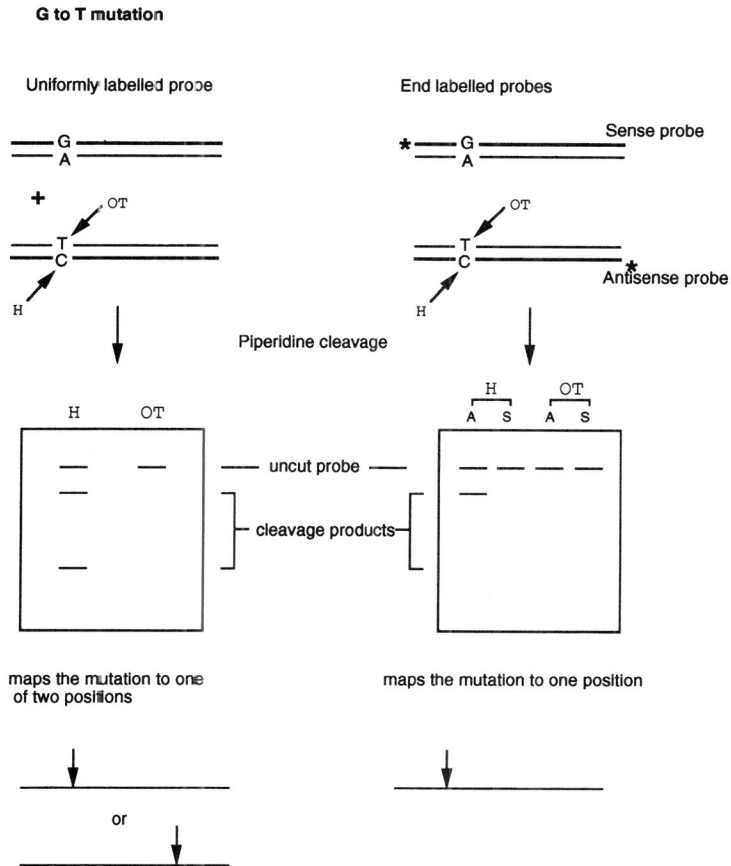

G to T mutation

Fig 5.16

Uniformly labelled probes map the mutation to one of two positions, whereas end-labelled probes map the mutation to a single position.

the position of the mutation. The pattern obtained by the cleavage will depend on how the probe was labelled. Conventionally, the probe is either labelled uniformly during the PCR amplification of the wild-type fragment or by end-labelling of the PCR primers. More precise map information is provided by using strand-specific end-labelled probes.

As can be seen in the example shown in *Figure 5.16*, the antisense strand specific end-labelled probe not only maps the mutation but also partially defines the nature of the mutation. In the example, a cleavage product is seen only in the hydroxylamine treated sample. Use of the reciprocal sense strand probe would yield no cleavage products in either the hydroxylamine or osmium tetroxide treated samples. Such a pattern of cleavage indicates that the mutation involves the substitution of a G base. The use of fluorescent dyes can permit the definition of the exact nature of the mutation (see Section 3.5).

3.2 Probing strategies

Typically both strands of a wild-type sample are labelled to produce the probe. If the probe is uniformly labelled, the sense probe and antisense probe will be used together. If the PCR primers are end-labelled, it is possible to generate a probe with one or both strands labelled. Using a single end-labelled strand, the mutation will be mapped to a single position, but this necessitates twice the amount of work as the other strand has to be handled separately. Given the extra work and that the DNA indicated to be mutant will be sequenced anyway, most workers choose not to use end-labelled strands separately.

To be sure of detecting insertions and deletions (and to be sure of detecting T–G mismatches), it is desirable to not only probe with both wild-type DNA strands, but also to probe the wild-type with both strands of the sample DNA (Forrest *et al.* 1991). Although extra work is required to make the probes, the probes can be mixed in equimolar amounts, such that both the wild-type and sample probes are present in a single tube, so the number of CCM reaction manipulations is not increased.

The CCM method is best suited to the analysis of long stretches of DNA. When short lengths are to be examined, the SSCP and DGGE methods are generally preferred as they are easier to perform. It is therefore likely that if you are considering CCM, you have a long stretch of DNA to test. Any length of DNA much above 1 kb, will require the use of multiple probes. To avoid mutations being missed by lying under a PCR primer or due to 'breathing' of the end of the duplex, probes should be overlapping. As a general guide, probes should be 800 to 1200 bp in length. This is a compromise between analysing a large piece (hence minimizing the amount of work required) and the resolving power of the gel. With probes in this size range, the region of overlap should be between 100 and 150 bp.

◇ The length of the probe is only limited by the resolution limit of the gel. Probes up to 2 kb have been used, but are not advisable as a mutation near the end of the probe may not be resolvable.

3.2.1 Heteroduplex formation

The heteroduplex is typically formed by mixing equal amounts of the sample and wild-type PCR products, followed by denaturation and re-annealing of the DNA strands. Half of the resulting duplexes should be heteroduplexes, and a quarter each homoduplex. The target DNA and the probe DNA should be identical in size.

It has been shown that heteroduplexes can form in the PCR amplification (when the mutation is heterozygous), thus preventing the need to generate the heteroduplex following the PCR. This presumably occurs during the final cycles of the PCR, when the amount of product is high and the levels of nucleotides, primers, and functional enzyme are low. Even if you are expecting a heterozygous mutation, it is advisable to generate heteroduplexes in the conventional manner, as you will not know how efficiently the heteroduplexes were formed in the PCR.

You should always include a negative control homoduplex sample. This sample should consist of probe-only duplex (i.e. no target DNA). This will serve as a background determinant. When setting up the CCM technique, it is also essential to include a positive control as a check for reaction conditions. It may not always be possible to obtain a positive control, e.g. when screening for mutations in a 'new' gene. It is imperative though, that you have a positive control when first setting up the technique.

The target DNA should be a clean PCR product, i.e. free of other bands, and be purified to remove PCR primers. The concentration of the template and probe DNA can be determined by running an aliquot on a gel alongside a known amount of marker DNA.

3.2.2 Determination of zygosity

As cleavage of the probe is never complete, it is not possible to readily know if the mutation detected is heterozygous or homozygous. In most cases though, you will have a good idea before you start, whether mutations will be heterozygous or homozygous. Sequencing of the mutant region will not only characterize the mutation but also determine the zygosity. If cloned sequencing is performed, half of the clones should represent one allele if the mutation is heterozygous. To avoid the chance of just sequencing one allele, several clones should be sequenced. Biases in the PCR or the cloning may also skew the proportion of the mutant to wild-type clones, further emphasizing the need to sequence more than just a few clones.

◇ Five clones should give a >95 per cent probability of having both alleles.

Direct sequencing can also determine the zygosity of the mutation in a known position. Heterozygous point mutations will be observed by the presence of a band in two tracks (or if fluorescently sequencing, two peaks at one position) (see Section 7). Heterozygous deletions or insertions will be observed by unreadability of the gel beyond the site of the mutation.

CCM itself can be used to determine the zygosity of the mutation. The analysis is done by making the probe from the sample DNA and annealing it to an unlabelled aliquot of the same DNA sample. If the mutation is heterozygous, cleavage will occur, but not if the mutation is homozygous.

3.2.3 The fume hood

In order to perform CCM analysis, you must have a large fume hood which can accommodate a radioactive-work facility. The labelling of the probe and generation of the heteroduplex stages of the method can be performed in an ordinary radioactivity area.

The preparation of the solutions should be performed with care in the fume hood. The modification reactions should be performed in the fume hood with radioactivity screens.

In the fume hood, you will require water baths or heating blocks for incubations. You will also require a microcentrifuge and bin for disposal of tips and tubes.

The supernatants of the ethanol precipitations after osmium tetroxide modification should be collected separately for safe disposal. After the ethanol precipitation following the piperidine cleavage, it is safe to handle the tubes outside of the hood.

3.3 Methodology

3.3.1 Probe preparation

5′ End-labelling

Usually oligonucleotides are manufactured with a free 5′-hydroxyl group. This allows the oligo to be end-labelled by the transfer of [^{32}P] phosphate from [γ-^{32}P]dATP using T4 polynucleotide kinase. After the labelling reaction, unincorporated hot nucleotides should be removed by spin column chromatography through a separation gel such as Bio-Rad Bio-Gel P4 Fine.

An aliquot of the labelled oligo should be counted on a scintillation counter to determine specific activity and an aliquot quantitated on a spectrophotometer. The labelled oligo is then suitable as a primer for PCR amplification to produce the CCM probe. The PCR product (the probe) when made, should be purified in the same way as a uniformly labelled probe (see below). In general end-labelled probes give less background than uniformly labelled probes.

Uniformly labelled probes

Uniformly labelled probes are generated by PCR amplification in the presence of a labelled nucleotide, usually [α-^{32}P]dCTP. A reduction in the level of cold dCTP helps the resultant probe to have a high specific activity. This has to be balanced with the problem of a high level of PCR-misincorporations caused by low concentration of a particular nucleotide. If the concentration of cold dCTP is greatly reduced, e.g. from 200 μM to 6 μM, the number of amplification cycles should be kept low, e.g. 25 cycles. If the dCTP concentration is slightly reduced, e.g. from 200 μM to 100 μM, a more usual number of cycles may be used, e.g. 32–35.

◇ Gel purification is often the method used for probe purification in CCM protocols. However, cutting a radioactive band out of a gel can be unpleasant and can cause contamination problems unless great care is taken.

The probe should be purified from unincorporated label by either spin column chromatography, ethanol/ammonium acetate precipitation (only larger molecules are pelleted), or gel purification.

Generating the heteroduplex

The probe/target heteroduplex is created by mixing the PCR product and probe, heating to 100 °C, and cooling. After re-annealing, the tube will contain mostly (unlabelled) target homoduplexes (as the target is in excess) and hot target/probe heteroduplexes.

If your gene of interest has a high G+C content (i.e. has a T_m higher than about 65 °C), it may be necessary to carry out the denaturation in a different buffer. The denaturation can be carried out in TE buffer (pH 8.0) and the annealing buffer can be added to a final concentration of 1× before transferring to the annealing temperature. The heteroduplex DNA is next divided into two for the modification reactions.

3.3.2 Preparing the solutions

The hydroxylamine solution

◇ Diethylamine is extremely irritating to the eyes and respiratory system. Use in a fume hood. It is also extremely flammable, so keep away from ignition sources.

Hydroxylamine hydrochloride or hydroxylammonium chloride should be dissolved in warm water, and the pH adjusted to pH 6.0 with concentrated diethylamine.

◇ Avoiding putting a potentially dirty pH probe directly into the hydroxylamine solution, will maintain the purity of the solution.

The pH of the stock solution should not be measured directly with a pH meter probe. It is safe however to measure the pH of a diluted aliquot. This should be done by removing 2 drops of the stock with a Pasteur pipette to 2 ml water in a bijou tube. The diethylamine should be added to the stock in 50 μl steps until the pH is correct. The hydroxylamine stock should be stored at 4 °C for up to 10 days.

The osmium tetroxide solution

◇ HEALTH WARNING! Osmium tetroxide causes burns to eyes and skin. The vapour is irritating to lungs, and can cause disturbance of vision. It is also a suspected teratogen. On the positive side, osmium tetroxide has a very pungent odour, so you know if you are being exposed.

Osmium tetroxide must be handled in a fume hood. It is best obtained as a 4 per cent aqueous solution in glass ampoules and should be of the highest purity available. When opened, the contents should be transferred to a small glass well-sealed container, as the solution will react with plastic over time. Once opened, the solution should only be kept for up to 3 months at 4 °C. Loss of potency is indicated by the solution taking on a greenish colour. The shelf-life of an aliquot may be extended by driving oxygen out of the vial by filling with nitrogen. It has been suggested that the solution may be storable for much longer by freezing aliquots at –80 °C.

◇ Pyridine, which is part of the osmium tetroxide buffer is a suspected teratogen. Avoid contact or inhalation. Thankfully pyridine also has an odour.

Should any osmium tetroxide be spilled, it should be wiped up with absorbent paper tissues and thoroughly wiped down with wet tissues. The used tissue, should be stored in a sealed glass container until disposal arrangements have been made.

Piperidine

◇ Piperidine is toxic and is a suspected teratogen. It also has a pungent odour.

Piperidine can be bought as a 10 M stock and should be diluted to a 1 M stock before use.

3.3.3 The CCM procedure

Hydroxylamine modification

Half of the duplex DNA should be mixed with hydroxylamine solution. If crystals have developed in the hydroxylamine solution, they can be dissolved by incubation at 37 °C.

◇ A typical hydroxylamine reaction time is 20 minutes.

Until you have experience of any particular probe, you will have to do some pilot work to establish the best reaction conditions, i.e. conduct reactions stopped at the time points 0 min, 30 min, 1 h, 2 h and 3 h. The reaction is stopped by the addition of 'stop' solution and ice cold ethanol for precipitation of the DNA. The stop solution contains tRNA which competes for the hydroxylamine, effectively stopping the modification reaction of the DNA, (and also aids precipitation of the DNA). The zero time point in the pilot study should be achieved by adding the stop solution to the DNA before the hydroxylamine solution.

Osmium tetroxide modification

◇ A typical osmium tetroxide reaction time is 5 minutes.

The other half of the duplex DNA should be mixed by pipetting on ice with osmium tetroxide buffer and osmium tetroxide. The solution should take on an intense yellow colour. A fading in the intensity of the colour is an indication of loss of the potency of the osmium tetroxide. Again, in initial experiments, the reaction will require optimization. Reaction times of 0 min, 1 min, 5 min, and 20 min should be performed. The stopping of the reaction is performed as above.

Piperidine cleavage

◇ Osmium tetroxide tubes may darken due to reaction with the plastic.

Following ethanol precipitation, the pelleted DNA should be dried and resuspended in 1 M piperidine. Cleavage is carried out at 90 °C for 30 minutes and the tube is then transferred to ice. If using a heating block, put a hot block on top of the tubes. This minimizes condensation on the lids of the tubes and also prevents the tubes from popping open. The samples should again be ethanol precipitated, dried, and dissolved in formamide loading dye (see Section 1.2). Tubes with fewer counts or more counts than average (as judged by a Geiger counter) should be dissolved in an according volume. The final samples should have retained about half the counts of the input DNA.

Preparing the gel

The gel used for CCM fragment analysis is identical to that used for DNA sequencing, i.e. a denaturing polyacrylamide gel. The apparatus, gel pouring, and disassembly of the apparatus is the same as that employed in SSCP analysis (see Section 1.2).

The denaturant is urea and this is added to the gel mix to a final concentration of 7 M. Additional denaturation is provided by the temperature of the gel, which should rise to 50 °C or more during the run. The acrylamide percentage will vary according to the size of the probe: 4 per cent gels should be used for probes greater than 1 kb, 5 per cent for probes between 500 bp and 1 kb, and 8 per cent for probes less than 500 bp. The gel should have the conventional acrylamide/bis ratio, i.e. 19:1, and contain TBE buffer at a 1× concentration. The gel should be pre-run for approximately 30 minutes prior to loading of the samples to warm it to above 40 °C.

Size markers

Because you will need to know the size of any cleavage products, it is necessary to load size markers on the gel. In order to be visualized on the autoradiograph, the markers have to be radiolabelled. There are many ways to make the markers and label them, but the simplest is to digest a plasmid with a restriction endonuclease which leaves a 5′ overhang at the cut site. Labelling is carried out by incorporating an $[\alpha\text{-}^{32}P]$, $[\alpha\text{-}^{33}P]$, or $[\alpha\text{-}^{35}S]$ dNTP into the overhang.

Enzymes such as *Msp*I and *Taq*I leave a 5′ overhang of which the 3′-most unpaired base is a G. The labelling can then be carried out by direct incorporation of labelled dCTP with Klenow polymerase or reverse transcriptase. The size markers should provide a roughly evenly spaced ladder from around the site of the probe to the bottom of the gel.

Running the gel

Prior to loading the gel, the wells should be flushed out with a Pasteur pipette to remove any urea which may have leached out of the gel. The samples should be placed in a boiling water bath to denature and then loaded onto the gel. The volume of the sample which can be loaded is dependent upon the type of comb used. Sharkstooth combs can only accommodate about 2 μl, whereas conventional combs can accommodate much more.

The gel should be run at or near the maximum Wattage, such that the gel is hot to the touch, i.e. around 50 °C. The bromophenol blue (dark blue) dye should be run to the bottom of the gel. Running times vary between 2 and 5 hours.

After electrophoresis, the gel apparatus should be dismantled as described in Section 1.2.2. If the probe was ^{32}P-labelled, the gel may be quite 'hot', so radioactive precautions should be taken. Prior to drying, the gel should be 'fixed'. This involves soaking the gel and the plate to which it is attached in 10 per cent methanol/10 per cent acetic acid for 20 minutes. Care should be taken when dunking the plate and when removing it from the solution, so that the gel does not become detached from the plate.

The gel should then be dried and autoradiographed. Auto-radiographic exposure of ^{32}P gels may be accelerated by the use of intensifying screens in the autoradiography cassette. The screens work by emitting light upon β-particle bombardment, thus intensifying the image on the film. The cassettes should be deposited in a –80 °C freezer, as the screens do not work at higher temperatures. An initial overnight exposure should be performed, followed by longer exposures if necessary.

3.4 The result

An ideal result is illustrated in *Figure 5.17*. In this case strand-specific probes were used. Cleavage is observed in the hydroxylamine-treated antisense probe only. This predicts the substitution of G. Sequencing of this region revealed a G to A mutation.

On developing the autoradiograph, you should first compare the sample tracks with the wild-type homoduplex negative control and the positive control (if you have one). The negative control will allow you to distinguish between what is background and what are cleavage products. The cleavage product is observed as a band not present in the wild-type homoduplex tracks. The presence of a positive control

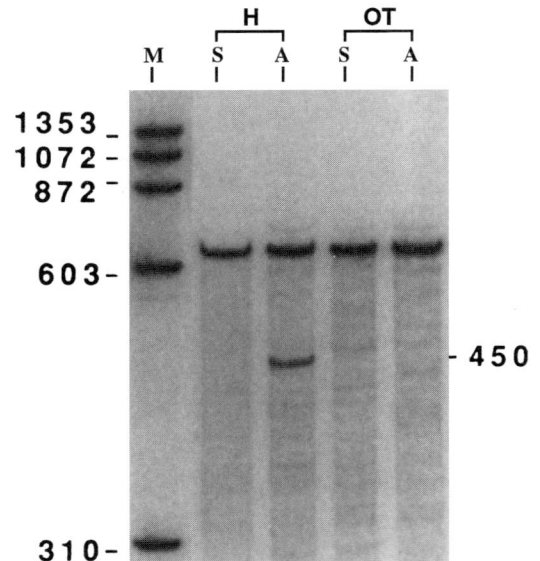

Fig 5.17

A point mutation (G to A) detected by CCM. M denotes marker; S denotes sense probe; A denotes antisense probe; H denotes hyroxylamine reactions; OT denotes osmium tetroxide reactions. (Reproduced with permission from Grompe *et al.* 1989).

will determine the extent of chemical modification and piperidine cleavage. It is essential that this sample looks good.

3.4.1 Problems

If you observe no cleavage in the positive control, it is likely that there was too little modification, too much target DNA, or under-reaction by piperidine. If you observe a ladder of bands it is likely there was too little target DNA or over-reaction by piperidine. If the probe band is absent, it is likely that the modification reactions were over-done. Under- or over-reaction can be corrected for by a corresponding alteration in reaction incubation time.

3.5 Application of fluorescence technology to CCM

CCM is readily adaptable to fluorescence technology (Haris *et al.* 1994; Verpy *et al.* 1994). Primers should be labelled as described in Section 1.4. The cost of dye-primers for CCM analysis is not as much of a concern as it is in SSCP analysis as fewer primers are required per kb of DNA.

Apart from the elimination of radioactivity, the great advantage of fluorescent labelling is that a greater number of samples can be handled more easily. It is possible to label one probe with one dye (labelling one or both strands) and mix four probes from four regions in one tube, thus greatly reducing the number of manipulations required. All four probes can be analysed in a single lane of the gel.

Alternatively, it is possible to label each of the wild-type and sample DNA strands with four different dyes, thus maximizing the mutation detectability. Using this approach, it is also possible to determine the nature of the mutation without sequencing. If the sizing of the cleavage products on the gel is very accurate, it should be possible also to determine which base has been mutated, i.e. it is theoretically possible to fully characterize a mutation without sequencing.

Solid-phase DNA technology can be applied to CCM analysis, and when coupled with fluorescence technology can greatly increase the number of samples that can easily be handled, and at the same time simplifies the sample manipulations (Rowley *et al.* 1995).

Table 5.2 The different cleavage patterns obtained from a single region using four probes[a], each labelled with a different dye (each type of mismatch gives a unique characteristic cleavage pattern[b])

	Hydroxylamine				Osmium tetroxide			
	mS	wtA	wtS	mA	mS	wtA	wtS	mA
A→T					+	+		
A→G				+		(+)		
A→C	+					+		
G→A		+						(+)
G→T		+			+			
G→C	+	+						
C→A			+					+
C→T			+		(+)			
C→G			+	+				
T→A							+	+
T→G				+			+	
T→C	−						(+)	

[a] Probe abbreviations: m, mutant; wt, wild-type; s, sense strand; A, antisense strand.
[b] +, cleavage; (+), osmium tetroxide modification of T–G mismatch.

◇ Advantages
● >95 per cent detection
● Predictable
● Nonradioactive

◇ Disadvantages
● Expensive PCR primers
● Time-consuming fragment design

4. Denaturing gradient gel electrophoresis

Denaturing gradient gel electrophoresis (DGGE) is a gel system which allows electrophoretic separation of DNA fragments differing in sequence by as little as 1 base pair. The separation is based upon differences in the temperature of strand dissociation of the wild-type and mutant molecules. As the fragments migrate down the gel, they are exposed to an increasing concentration of denaturant in the gel. When the molecules reach a critical denaturant level, the DNA strands begin to dissociate. This causes a significant reduction in the fragment's mobility. As the position of this critical point is a function of the melting temperature, the point at which mobility retardation occurs for wild-type and mutant molecules will be different, thus resulting in their separation.

DGGE does not require radioisotopes or toxic chemicals, but does require some specialist equipment. The mutation detection rate is close to 100 per cent.

The main difference between DGGE and other mutation scanning techniques is that the behaviour of DNA molecules on DGGE gels can be modelled by computer, hence that the detectability of a mutation in a given fragment can be accurately predicted.

Once the equipment is in place and the experiments have been designed, the technique is relatively simple to perform.

Fragment sizes are limited to between 100 and 800 bp due to the resolution limit of polyacrylamide gels.

4.1 Theory of DGGE

Melting (or denaturation) of DNA is the dissociation from a double-stranded helical conformation to a single-stranded non-helical conformation. DNA can be melted by high temperature or by chemicals. The melting temperature (T_m), the temperature at which strand dissociation occurs, is determined by the sequence composition of the DNA fragment. The forces holding a DNA helix together are not just provided by the hydrogen bonding between purine and pyrimidine base pairs, but also, and to a larger extent, by stacking interactions between adjacent bases on the same strand. Thus it is the actual sequence of a particular DNA fragment, not just the GC content which determines the T_m. Thus a G to C mutation, which does not alter the GC content, is just as detectable as other mutations.

Within a fragment of DNA, there may exist domains of different melting temperature. These melting domains vary in length between 25 and several hundred base pairs and the melting temperatures typically vary between 65 and 80 °C.

The DGGE system employs a gradient of denaturant in the gel (provided by the chemicals urea and formamide), so that the further a fragment migrates through the gel, the higher the level of denaturant it encounters. At a certain point in the gel, a critical level of denaturant will be reached corresponding to the T_m of the domain, forcing the DNA strands to separate. The partially melted molecule becomes branched in structure and travels through the pores of the gel with much more difficulty. Until the DNA fragments begin to denature, their mobility is dependent on their size. When a fragment carries a base substitution, the entire domain in which the variant base resides will have an altered T_m. Thus the wild-type and mutant fragments will begin to denature at different points in the gel and thus become separated (*Figure 5.18*) (Fischer and Lerman 1983). A particular fragment may have several melting domains. Mutations are detectable in any of the domains except the highest melting domain, as when this domain melts, complete strand dissociation occurs, so that the fragment is no longer branched and once again migrates according to its size.

The problem of nondetection of mutations in highest melting domains has been solved by the use of 'GC-clamps' (see Section 4.1.2). These are regions of very high GC-content DNA which are attached to (usually) one end of the fragment to provide a new synthetic highest melting domain.

DGGE allows the detection of mutations in both heteroduplex DNA and homoduplex DNA. Mutant and wild-type homoduplexes can

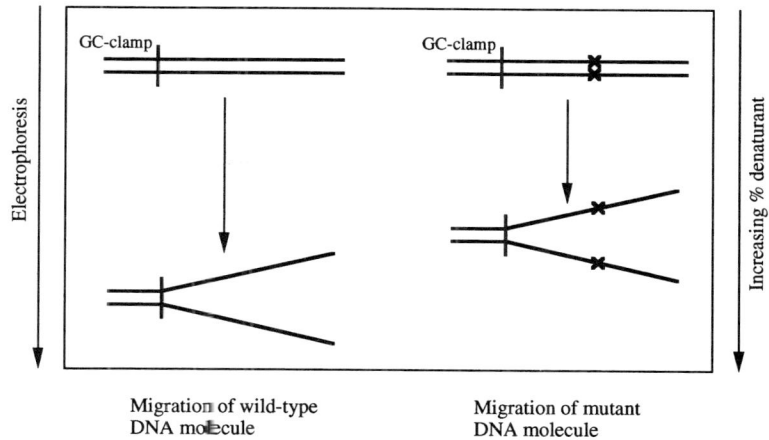

Fig 5.18

Diagrammatic representation of DNA fragment behaviour through a denaturing gradient gel.

usually be separated, but the detection rate and resolution of hetero-duplexes is significantly better as a greater loss of stacking forces occurs around the mismatch. In some cases, the stacking forces in the vicinity of a mismatch are sufficiently changed that an alteration occurs in the melting domain structure. Under optimal conditions, heteroduplexes can virtually always be separated.

Although DGGE is normally used to analyse DNA fragments amplified by PCR, the technique is also usable for the analysis of genomic DNA. In this case, the gels are electroblotted and probed.

4.1.1 DGGE gel systems

There are three types of gel system used in DGGE: perpendicular and parallel denaturing gradient gels and constant denaturing gels. Parallel gels are the most commonly used form of denaturing gradient gel.

Perpendicular denaturing gradient gels

In perpendicular DGGE, the denaturant concentration increases from one side of the gel to the other. The DNA is loaded into a single long well, which lies across the top of the gel. This results in a very peculiar looking gel. Molecules on the less denaturing side of the gel are not melted and therefore migrate according to their size. Molecules migrating on the more denaturing side of the gel are partially melted and therefore migrate slowly. Molecules migrating in the strongest concentration of denaturant may even undergo complete strand dissociation. These molecules are not branched and therefore migrate faster than partially melted molecules. For a fragment consisting of two domains (the higher melting domain is usually a GC-clamp) there will be a steeply sloping region of DNA between the fully helical and partially melted molecules (*Figure 5.19*). This region corresponds to the denaturant concentration at which both forms are present at a significant level. The relative levels of the two forms alter significantly at only slightly different denaturant concentrations. It is at or around this concentration of denaturant that molecules differing by a single base pair can be distinguished. A

Fig 5.19

Diagrammatic representation of a perpendicular denaturing gradient gel. The DNA fragments are seen as 'S-shaped' bands. This example consists of wild-type and mutant homoduplexes. If heteroduplexes were also included, a total of four bands would be seen. The mutant homoduplex here, has a slightly higher T_m than the wild-type. The greatest separation of the two forms occurs in this example between 40 per cent and 65 per cent denaturant. Typically, the range of denaturant at the steepest part of the slope varies around 25–30 per cent.

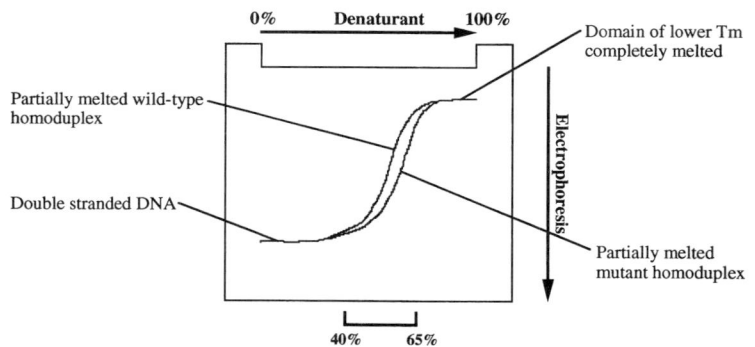

substitution which lowers the T_m of the fragment, will cause the sloping region to be positioned slightly towards the lower denaturant side of the gel. Reciprocally, a substitution resulting in a higher T_m, will shift the slope towards the higher denaturant side of the gel. In either case the mobilities of the purely helical DNA molecules or molecules in which the lower melting domain is completely melted are not distinguishable between the mutant and wild-type.

Perpendicular gels will detect all mutations on a given fragment but are limited in their use as they can only examine one sample at a time. The power of perpendicular DGGE however, is to provide an experimental guide to the conditions for parallel DGGE.

Parallel denaturing gradient gels

In parallel DGGE, the denaturant increases from the top of the gel to the bottom, i.e. the gradient is in the same orientation as fragment migration. These gels are more conventional in their appearance, as normal gel combs are used, allowing the analysis of multiple samples, and the fragments migrate as normal-looking bands.

Unlike the perpendicular gradient gels which generally have a 0 to 100 per cent denaturant range, the parallel gradient gels use a smaller denaturant range, corresponding to the steepest part of the curve on the perpendicular gradient gel. This narrow range of denaturant concentration allows better separation of fragments.

The fragments migrate as helical molecules until the denaturant concentration is sufficient to cause melting, and so mobility reduction (*Figure 5.20(A)*). For molecules differing by a single base pair, the denaturant concentration at which mobility is retarded will differ from the wild-type. In most cases of mutation detection (i.e. heterozygous mutations), heteroduplex molecules will be formed during the PCR reaction. The heteroduplex bands melt at a lower denaturant concentration and produce much greater 'band-shifts' than the homoduplex bands (*Figure 5.20(B)*). After the strands in the lower melting domain have separated, the mutant and wild-type fragments continue to migrate, but at a dramatically slowed, and approximately equal, rate. Complete strand dissociation rarely occurs, as very long run times are required to get the slow moving fragments into a region of sufficiently high denaturant.

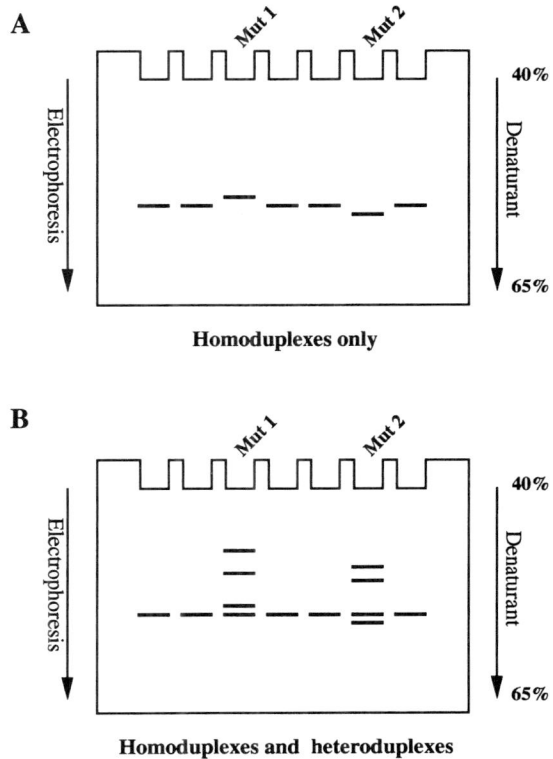

Fig 5.20

Diagrammatic representation of parallel denaturing gradient gels. The denaturant range is determined by the perpendicular gradient gel (*Figure 5.19*). A, Homoduplex DNA PCR amplified from a haploid organism. Two mutants are visualized as band shifts. B, DNA PCR amplified from a diploid organism showing two heterozygous mutations. The heteroduplex bands show greater retardation of migration than the homoduplex bands.

If more than one domain is to be examined on a fragment, multiple gels with different gradient ranges may have to be run.

Constant denaturing gels

Constant denaturing gels are the simplest form of the DGGE mutation detection system. The gradient is eliminated, as the gel contains only a constant concentration of denaturant which coincides with the T_m of the domain of interest. It is essential therefore that the denaturant concentration is precisely optimized for each domain. CDGE allows increased resolution as variant fragments have different mobilities, but these mobilities are unchanged during the run.

The sensitivity of CDGE is very high and can be used for the detection of mosaic oncogene mutations in tissue down to the 5 per cent level. Although sensitive, CDGE is not a valid alternative for mutation scanning over a large region, as each domain has to be precisely optimized. It is most useful when you are examining many samples for mutations in a single domain (Hovig *et al.* 1991).

Temperature gradient gel electrophoresis (TGGE)

A variant on the denaturing gradient gel concept is TGGE (Wartell *et al.* 1990). In this version, the gel contains a homogeneous concentration of denaturant, and the gradient is provided by a temperature difference from one side of the gel to the other (perpendicular TGGE) or from the top to the bottom of the gel (parallel TGGE).

The gradient is provided by water of different temperatures flowing through channels in the sides or top and bottom of aluminium plates which are attached to the inner and outer plates of the gel assembly. Although this system eliminates the need for pouring gradient gels, it requires more specialist equipment and therefore has not been widely adopted.

4.1.2 GC clamps and chemical clamps

GC clamps

GC clamping is a development which provides an artificial highest melting domain (Sheffield *et al.* 1989). The clamp is a run of only Gs and Cs, usually of about 40 bases in length. The clamp is typically attached to the target fragment by incorporating the clamp sequence onto the 5′ end of one of the PCR primers (*Figure 5.21*). This primer tail gets copied in the amplification reaction, providing a double-stranded GC-rich region at one end of the PCR product.

The clamp sequence does not appear to interfere significantly with the amplification process. The positioning of the clamp at the left or right of the intended fragment may often not matter, but will sometimes be better on one particular side (see Section 4.1.3). The GC-rich primer should be stored in aliquots as freezing and thawing may lead to degradation of the clamp. The effect of the GC clamp is to render all base changes in the amplified fragment detectable by DGGE. The disadvantage of the clamp is its cost. One half of all the PCR primers used in a DGGE mutation screening experiment will be approximately 60 bases in length, thus making the experiment expensive.

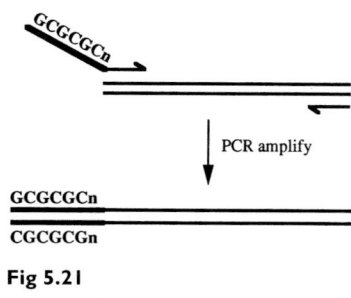

Fig 5.21

Addition of the GC-clamp.

Chemical clamps

Psoralen-modified PCR primers provide an alternative to GC clamps. Psoralens are photoreagents that form covalent bonds with pyrimidine bases (mainly thymidines). In this procedure (Costes *et al.* 1993), 5-(ω-hexyloxy)-psoralen is attached to two successive As at the 5′ end of an oligonucleotide during its synthesis. When used as a PCR primer, the psoralen derivative becomes intercalated between the two DNA strands (*Figure 5.22*). The PCR product is then UV-irradiated, covalently linking the two strands.

◇ Psoralen-linked oligonucleotides are commercially available from Appligene.

The crosslinking reaction, known as ChemiClamping is very efficient. The clamp behaves in much the same way as a GC clamp but is perhaps a little superior in that the clamp is stronger. A disadvantage of the chemical clamp however, is that the bands on the gel are less intense than the GC-clamped bands.

4.1.3 DGGE fragment design

◇ MELT and SQHTX are available on request from Dr L. S. Lerman; e-mail: lerman@fang.mit.edu

◇ MacMELT™ is a product of MedProbe, Oslo, Norway.

Before embarking on a series of DGGE experiments, you must design the fragments which will cover the region to be analysed. At its simplest, the fragments consist of a single melting domain attached to a GC clamp. The programs MELT and MacMELT assist fragment design and the program SQHTX assists optimization of experimental conditions (Lerman and Silverstein 1987).

Fig 5.22

Addition of the chemical clamp.

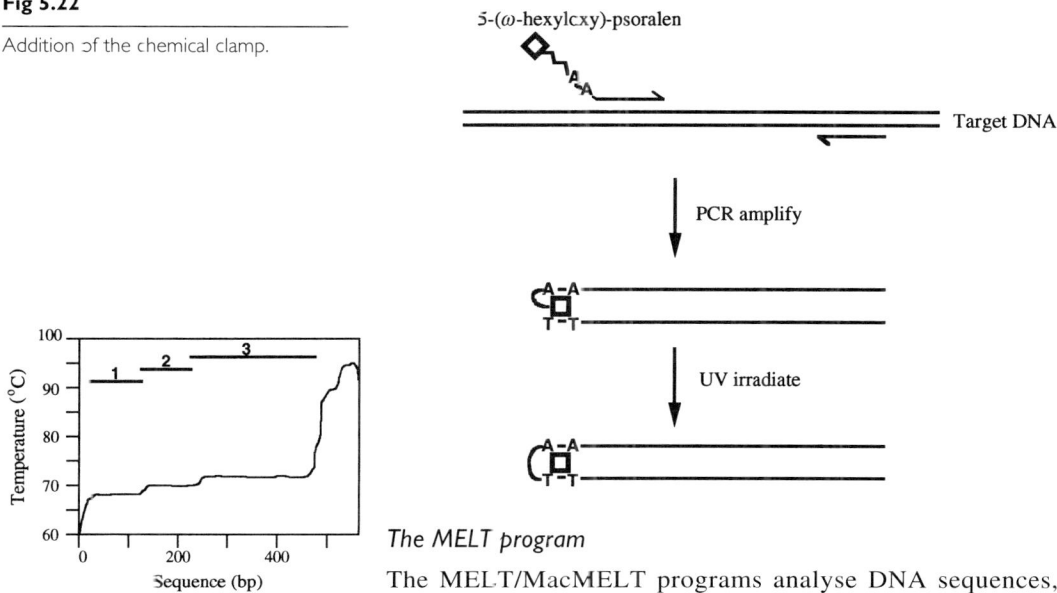

5-(ω-hexyloxy)-psoralen

Target DNA

PCR amplify

UV irradiate

Fig 5.23

A MELT profile of a DNA fragment with three melting domains.

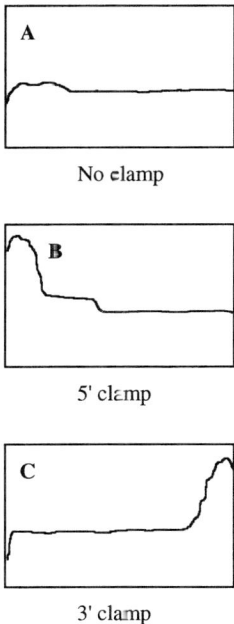

A

No clamp

B

5' clamp

C

3' clamp

Fig 5.24

MELT maps of a DNA sequence. A, with no GC-clamp; B, with a GC-clamp on the 5' end; C, with a GC-clamp on the 3' end. Positioning the clamp on the 3' end in this case converts the DNA fragment into a single melting domain.

The MELT program

The MELT/MacMELT programs analyse DNA sequences, modelling the melting profile and graphing the data. Experimental evidence indicates that the computer-simulated melting profile is indeed accurate.

A melt map is a plot of the temperature at which each base pair of a fragment is in 50:50 equilibrium between the helical and melted forms. At temperatures below the line, a base pair will be helical and at temperatures above the line, it will be melted. Although the melt map is plotted in terms of temperature, there is a direct linear relationship between this and denaturant concentration in the gel.

Melting domains appear as horizontal lines on the melt map. The number and extent of domains are readily observable as discrete regions. The boundaries between domains are often quite abrupt.

When a fragment consists of multiple domains, the GC-clamp is usually put on the side of highest T_m. It is also desirable that the fragment is chosen such that the T_m increases from one side to the other, in 'staircase' fashion.

The melt maps in *Figure 5.24* illustrate a fragment consisting of two domains and the effect of a 40-bp GC clamp at the 5' and at the 3' end. Differences in the melt map when the clamp is positioned at the 5' or 3' end occur because the domain structure is a property of the entire fragment.

The SQHTX program

The program SQHTX indicates optimal denaturant conditions for the gel and the run time required. The program therefore provides an alternative to perpendicular gradient gels in the determination of running conditions for parallel gradient gels. The melting temperature as predicted by MELT, is converted into an equivalent chemical denaturant concentration. As a rule of thumb, for a gel run at 60 °C, a 1 per cent increase in denaturant corresponds to an approximately 0.3 per cent increase in the theoretical melting temperature. Therefore the melting temperature approximately equals $60 + 0.3x$, where x is the denaturant percentage. So for a domain with a T_m of 72 °C, the desired denaturant concentration $\approx 72 - 60/0.3 = 40$ per cent.

SQHTX calculates the effect on mobility of variants at each base pair position in the fragment for a series of gel running times. Thus SQHTX complements the MELT program by providing a quantitative estimate of the mobility of a branched molecule.

If SQHTX indicates that very long run times are needed for melting the domain of highest T_m, it may be preferable to run additional gels with a different denaturant range. In general, most people choose to avoid such problems by choosing fragments which either contain just one domain of interest or more than one domain but with similar T_ms so that all mutations can be detected on a single gel without the need for long run times.

4.2 Methodology

4.2.1 Fragment design

In planning a DGGE experiment you must first decide if you need to detect all mutations or just most. If you want maximum detection, you will have to invest some time in scrutinizing the DNA sequence with the MELT program. PCR allows the end points of fragments to be chosen with little restriction. Fragments with extremely high GC-contents may not be helped by GC-clamping. In such cases the T_m may be lowered by replacing dGTP with dITP in the PCR reaction. It is possible for some regions of very high GC-content, that DGGE would not be suitable for mutation detection.

Fragment size is limited to a maximum of about 800 bp, by the resolution limit of the polyacrylamide gel. In reality fragments will rarely be greater than 500 bp in size due to problems of multiple domains with sizeable T_m differences.

If the fragment sequence is not known but primers are available for PCR amplification, computer prediction of fragment behaviour will not be possible. In such a case, a perpendicular denaturing gradient gel should be run, using a denaturant range of 0–100 per cent. The number and properties of domains can be determined, and the information used to derive conditions for parallel DGGE.

4.2.2 Generating the heteroduplex

DGGE is very sensitive and it is therefore generally not necessary to specifically generate heteroduplexes. A sufficient level of heteroduplex is formed in the final cycles of the PCR amplification as molecules synthesized in previous cycles denature and re-anneal with each other. This, of course, assumes that anticipated mutations are heterozygous not homozygous. If the mutation is expected to be homozygous, a heteroduplex should be generated between the sample and wild-type PCR products.

4.2.3 Reagents

The acrylamide is best obtained as 40 per cent stock solution at 37.5:1 acrylamide: bisacrylamide. Zero per cent denaturant stock is acrylamide (usually at 8 per cent) in 1× TAE buffer; 100 per cent denaturant is acrylamide of the same concentration in 40 per cent formamide and 7 M urea (42 g/100 ml).

4.2.4 Preparing the Gel

A disadvantage of the denaturing gradient gel system is that the gels have to be run in an accurately temperature-controlled heated buffer tank and the gel gradients have to be made.

Specialist equipment for running perpendicular and parallel gels is available from CES Scientific (Del Mar, CA, USA) and Bio-Rad. The Bio-Rad system includes a simple gradient mixer and perpendicular gel pouring device. Many early DGGE users made their own gel running systems from components from various sources.

The gel pouring guide below is intended for nonspecialist equipment. Specialist equipment, e.g. Bio-Rad 'D-GENE™' comes with thorough instructions.

Methods of forming perpendicular gradients

The gel plates should be thoroughly cleaned as described in Section 1.2.1. There are several different ways of producing a perpendicular denaturing gradient gel. One way is to seal three sides of the plates with spacers and clamps and pour in the gel mix (*Figure 5.25*). A flat edge to the top of the gel is created by gently overlaying with

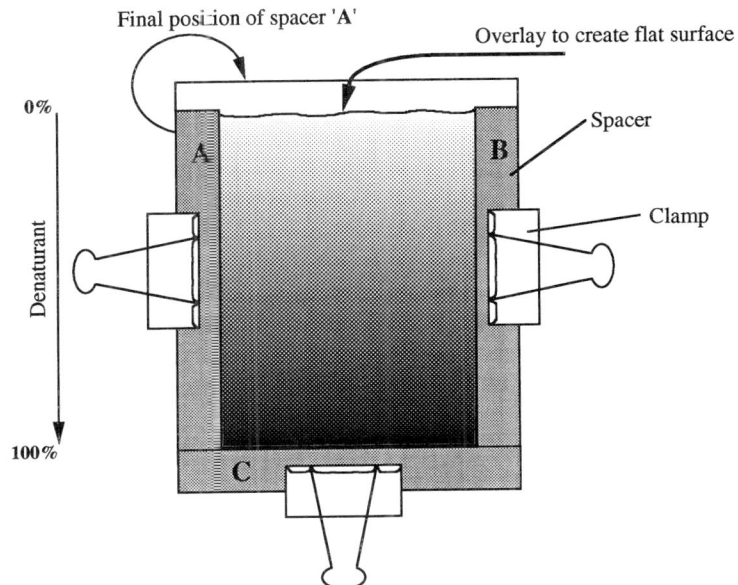

Fig 5.25

Diagrammatic representation of pouring a perpendicular gradient using the most basic components.

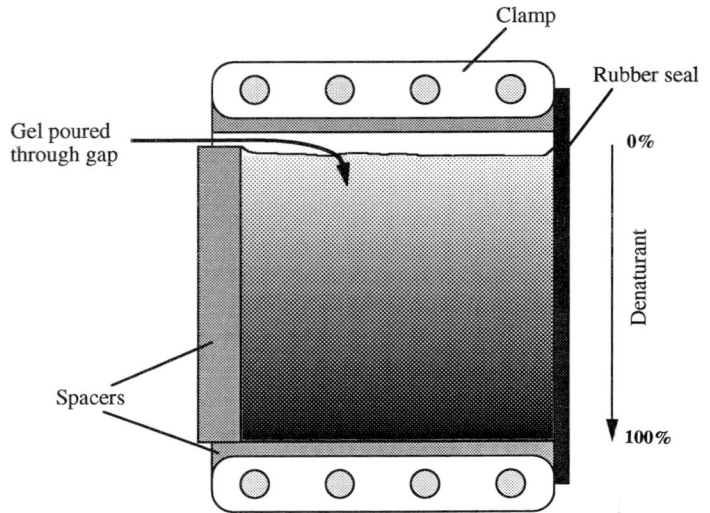

Fig 5.26

Diagrammatic representation of pouring a perpendicular gradient using a standard polyacrylamide gel unit.

0.1 per cent SDS or water-saturated butanol. When the gel has set, the SDS is flushed out with running buffer. The left-hand side spacer (A) is then removed and placed into position along the top of the gel. The right hand spacer (B) is then removed. The gap formed by the removal of this spacer forms the well. The entire assembly is then rotated 90° anticlockwise to produce a gel with an increasing gradient running from the left to the right. An additional spacer may be useful for extracting the spacers from the assembly.

An alternative way is to use a standard polyacrylamide gel unit, such as Hoefer or Bio-Rad mini-Protean II, which uses a rubber casket to seal the bottom and clamps to seal the sides (*Figure 5.26*). A shortened spacer is inserted into the top of the gel (partly protruding), leaving a small gap at one end. The assembly is then turned 90° so the gap is now at the top of one side. The gel is poured through the gap. When full, the gap is sealed with modelling clay. When set, the pro-

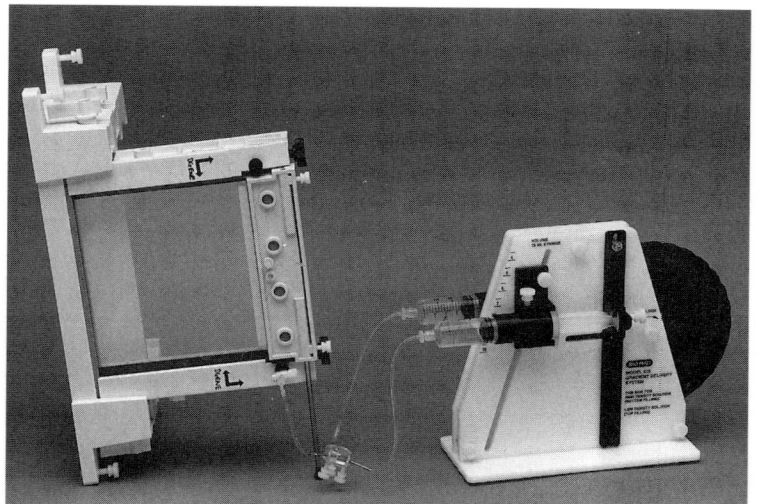

Fig 5.27

The Bio-Rad 'D-GENE' system for pouring perpendicular gradient gels.

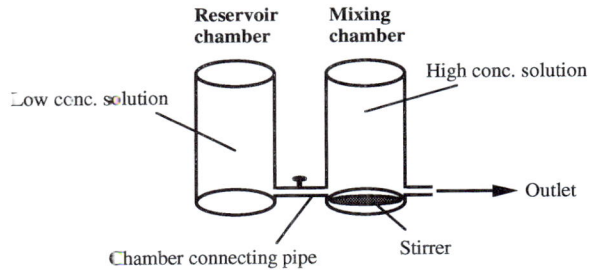

Fig 5.28

Diagrammatic representation of a simple gradient-maker.

truding spacer is removed, to produce a long flat well. If a gap of air exists at the side of the gel, it can be filled with molten 1.5 per cent agarose in running buffer or 0 per cent denaturing polyacrylamide.

All of these complications are avoided in the Bio-Rad system as the gel is poured through a gap in the spacers (*Figure 5.27*). This permits all four sides of the gel assembly to be in place when the gel is poured.

Spacers for perpendicular gels are generally 1 mm thick, but narrower spacers will give sharper bands.

Pouring a gradient gel

The acrylamide concentration will depend upon the size of the fragment to be analysed. An 8 per cent gel is appropriate for the resolution of fragments between 150 and 500 bp. For smaller or larger fragments, the acrylamide concentration should be altered accordingly.

The simplest form of gradient mixer consists of two chambers connected by a tube (*Figure 5.28*). The chamber containing the more highly concentrated solution, has a long outlet tube which fills the gel assembly. A propeller or magnetic stirrer is used to mix the solution in the more concentrated chamber (the mixing chamber). Mixing of the gel solution should not be so vigorous as to introduce bubbles, as the dissolved oxygen will inhibit polymerization.

Ammonium persulphate and TEMED (see Section 1.2.1) should be added to beakers of the two gel solutions and mixed by swirling. The more concentrated solution should be poured into the mixing chamber and the valve interconnecting the two chambers briefly opened to allow some gel solution to pass into the reservoir chamber. The solution which passed into the reservoir chamber should be transferred back to the mixing chamber. This process prevents bubbles blocking the interconnecting pipe. The reservoir chamber should then be filled with the less concentrated denaturing gel solution and the connecting valve gently opened.

The outlet should then be opened slowly, and the gel filled from the top. As the more concentrated solution in the mixing chamber flows out and into the gel assembly, the levels of the two chambers are kept even by flow of the less concentrated solution (in the reservoir chamber) into the mixing chamber. In this way the denaturant concentration goes from high to low as the gel pours. A peristaltic pump may be used to speed up and control the rate of flow. The filling tube is often a piece of Tygon flexible tubing with a micropipette tip attached

◇ Perpendicular gels typically have a 0–100 per cent or 20–80 per cent denaturant range. Shallower gradients (e.g. 10–60 per cent) that span the optimal concentration will increase the resolution of heteroduplex and homoduplex molecules.

◇ To be sure that the gradient is well formed, including a little bromophenol blue in the high-concentration solution will allow you to visualize the gradient by the intensity of the blue dye in the gel.

to the end. The tip should be positioned so that the gel flows down one plate only. 100 per cent denaturant will flow into the assembly first, followed by mix of decreasing concentration. The flow rate should be such that the gel takes about 5 minutes to pour. As the last of the gel mix is dispensed from the mixer, the flow rate should be reduced if possible, so that it doesn't plop out of the pouring tube disrupting the gradient.

In all examples of nonspecialist gel assemblies for perpendicular gradient gels, leakage will always be the main concern. The use of vacuum grease on the end of the spacers to seal the junctions will help prevent leakage. Pouring a little 1.5 per cent agarose into the system, can be used to seal the edges prior to pouring the gel. Gels can be poured in advance and stored at 4 °C for a few days.

Parallel denaturing gradient gels

The pouring of parallel gradient gels is much more straightforward than that of perpendicular gels. The gradient is generated in the same manner, but an ordinary polyacrylamide gel unit such as Hoefer or Bio-Rad is used. The two gel solutions, instead of being 0 and 100 per cent, should be of the concentrations as determined by perpendicular DGGE or the SQHTX program. The melting transition point should be approximately in the middle region of the gel. The comb should be inserted gently and at angle. This prevents disturbance of the gradient and the trapping of air bubbles under the comb. After the gel has set, the comb should be removed. This is best done by holding the gel assembly upside down and at a slight angle. This enables any unset liquid in the wells to drip out as the comb is withdrawn. The wells should then be flushed out with running buffer.

Constant gradient gels

The pouring of constant-gradient gels is done exactly as any conventional polyacrylamide gel. The gel can be poured relatively quickly (in about 1 minute). The comb should either be inserted after the gel has been poured (as above) or inserted before pouring, and the assembly held at an angle during pouring so that bubbles do not form under the wells.

4.2.5 Running the gel

Once polymerized, the gel should be immersed in a temperature controlled tank. The tank will usually be filled with running buffer (1× TAE), and may also contain the anode electrode. In order to maintain the pH of the buffer during the run, a peristaltic pump should be used to pump buffer from the tank up into the cathode buffer chamber of the gel apparatus. The excess buffer will overflow back into the tank. The tank should normally be accurately maintained at 60 °C. Fragments with a T_m of below 60 °C, have to be run at a lower temperature. Some DGGE users overset the tank temperature by about 2 °C prior to loading, as the tank can cool by this amount during gel loading.

After loading the tank is then reset to the correct temperature. To minimize evaporation the tank should be covered with a lid or polypropylene spheres. To prevent localized temperature differences, the buffer in the tank should be mixed with a propeller or magnetic stirrer. The gel should be pre-run for a short while to ensure that the gel is hot.

For perpendicular gels, the PCR amplification should be carried out in a 100 μl volume. The PCR product should be mixed with a 0.5 volume of loading buffer. Prior to loading, the wells should be flushed out with running buffer using a Pasteur pipette. The gel should be run at 60–150 V for the time determined by the SQHTX program or by past experience. It is important that the voltage is not so high that it heats the gel above the desired temperature.

For parallel gels, it is advisable to perform a time-course gel, in which a positive control sample is loaded repeatedly at time intervals to determine the optimal run time. Run times should be sufficiently long to get good fragment separation and sharp focusing of the bands. Perpendicular gels usually take 2–3 hours to run (the dye is run to the bottom) and parallel gels 2–16 hours. Certain rare sequences exhibit a succession of small domains extending over a large melting temperature range. Such a fragment may require a relatively long run at diminishing migration rate before a stable, final position is reached.

Staining and photographing the gel

When the run is complete, the power supply should be turned off and the gel assembly removed from the tank. Gels run overnight may require a timer to switch the power off to prevent over-running.

To separate the plates, the side clamps should be removed and then one of the spacers should be partially withdrawn and twisted. Gels of an acrylamide concentration of 8 per cent and higher can be picked up without breaking. Gels less than 8 per cent may break if picked up and should therefore be kept on the glass plate. The gel should be transferred to a dish containing 0.5 μg/ml ethidium bromide and gently mixed for 5–10 minutes. The gel should then be destained in water for about 5 minutes. This has the effect of reducing the background staining of the gel itself. The gel is then photographed on a UV transilluminator.

When photographing perpendicular gels, a ruler should be placed at the bottom of the gel, with 0 cm set at the low denaturant side of the gel. This aids the determination of the denaturant concentration at the steep part of the curve.

4.3 The result

An ideal parallel DGGE result is shown in *Figure 5.29*. Individuals of a homozygous genotype yield a single homoduplex band, whereas heterozygous individuals yield two homoduplex and two heteroduplex bands. From the banding pattern in this Figure, it is possible to determine the genotype of each individual.

◇ The loading buffer is 20 per cent Ficoll 400, 0.1 per cent bromophenol blue in 1× TAE (0.04 M Tris-acetate, pH 7.6, 0.001 M EDTA).

1 2 3 4 5 6 7 8 9 10 11 12

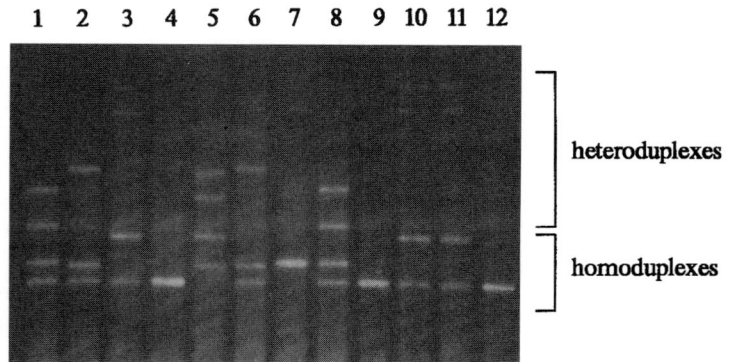

Fig 5.29

A parallel denaturing gradient gel. Twelve individuals are genotyped for three single base variants within a single DNA fragment, generating four possible haplotypes. Six different genotypes are displayed on this gel. (Reproduced with permission from Fodde and Losekoot, 1994; John Wiley & Sons, Inc.).

heteroduplexes

homoduplexes

As with any of the PCR-based mutation detection systems it is important that the PCR is optimized to produce a good yield of product without background.

5. Heteroduplex analysis

Heteroduplex molecules—double-stranded DNA molecules containing a mismatch—can be separated from homoduplex molecules on ordinary gels. The mutation detection rate of heteroduplex analysis is unknown, but it is clearly significantly lower than 100 per cent. It would appear that it is not the nature of the mismatch, but the sequence of DNA flanking the mismatch that affects the detectability. Mismatches in the middle of DNA fragments are detected most easily.

What heteroduplex analysis lacks in sensitivity, it makes up for in simplicity. The method requires no unpleasant chemicals, radio-isotopes, or specialist equipment.

◇ Advantages
● Simple
● Cheap

◇ Disadvantages
● Probable low detection efficiency

5.1 The theory

When double-stranded DNA derived from a mutant allele is mixed with corresponding wild-type DNA, and the fragments are heated to denature and cooled to re-anneal the strands, four types of molecule are produced. Half of the resultant molecules are identical to the input molecules (homoduplexes) and half are hybrid molecules (heteroduplexes) (*Figure 5.30*).

When run on a gel, the homoduplex bands migrate according to their length. The migration of heteroduplex bands, however, is not totally size-dependent (Nagamine *et al.* 1989; White *et al.* 1992). It is unclear exactly how the mobility of the heteroduplex molecules is altered, but it is most likely due to the production of a kink or mismatch bubble in the helix (*Figure 5.31*). A kink in the DNA, will cause the molecule to have a slowed mobility as it will pass through the pores of the gel matrix with greater difficulty. A mismatch may also cause a slight destabilization of the helix producing a more 'open' double stranded

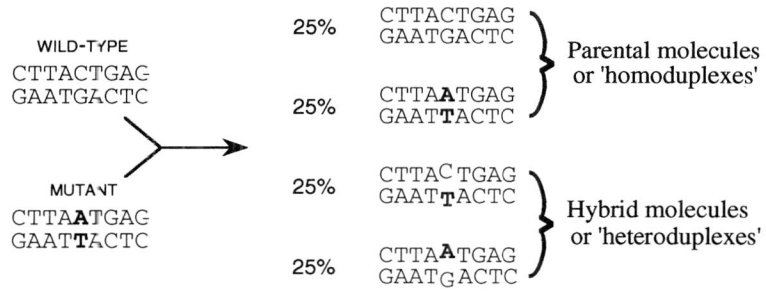

WILD-TYPE
CTTACTGAG
GAATGACTC

MUTANT
CTTA**A**TGAG
GAAT**T**ACTC

25%
CTTACTGAG
GAATGACTC

25%
CTTA**A**TGAG
GAAT**T**ACTC

} Parental molecules or 'homoduplexes'

25%
CTTAC TGAG
GAATT ACTC

25%
CTTA**A**TGAG
GAATG ACTC

} Hybrid molecules or 'heteroduplexes'

Fig 5.30

Heteroduplex molecules are double-stranded DNA molecules which contain a mismatch.

configuration surrounding the mismatch, which will also slow the migration of the molecule. This possibility is supported by the fact that the inclusion of a moderate amount of the denaturing chemical urea, appears to enhance the migration difference between homoduplexes and heteroduplexes, presumably by further loosening the helix.

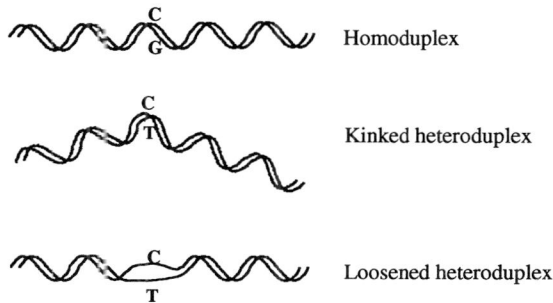

Homoduplex

Kinked heteroduplex

Loosened heteroduplex

Fig 5.31

Heteroduplex molecules may have retarded electrophoretic mobility because they are kinked or bulged at the point of mismatch.

Single base deletions or insertions are easier to detect than single base substitutions by heteroduplex analysis. Again this is probably due to loosening of the helix, which is greater when a base is deleted or inserted than when a mismatch occurs. There is no evidence yet to suggest that some types of base mismatch are easier to detect than others. It is therefore likely that all types of mismatch can be detected with equal probability. This, together with data showing that the same mismatch in different sequence contexts shows differing detectabilities, indicates that it is the DNA sequence flanking the mismatch which has the main influence on heteroduplex analysis detectability. As described in Section 4.1, it is apparent that an alteration in the helix stacking forces is responsible for the local context effect.

There is some evidence to suggest that mutations in the middle of a DNA fragment are most likely to be detected by heteroduplex analysis. This is probably due to the fact that a kink in this position would be expected to cause the greatest mobility change.

5.2 Practical considerations

Heteroduplex analysis consists of simply running a (heteroduplex) PCR product on a long polyacrylamide gel, and staining the gel to visualize the DNA fragments.

5.2.1 Gel type

The gel matrix used for heteroduplex analysis is either polyacrylamide or MDE gel (see Section 1.1.1). MDE gel has a better heteroduplex analysis mutation detection rate than polyacrylamide (Keen *et al.* 1991; Ravnik-Glavač *et al.* 1994*b*). MDE is a modified acrylamide-based vinyl polymer. It has a higher loading capacity than conventional polyacrylamide giving higher sensitivity without loss of resolution. When polyacrylamide is used, it is at 5 per cent C or 3.3 per cent C (see Section 1.1.1) and at acrylamide concentrations of 5–12 per cent, depending on fragment size. The gel apparatus used is usually a sequencing gel assembly and is generally about 20×40 cm with 0.8 or 1 mm-wide spacers. Glycerol is often included in heteroduplex analysis gels as it is in SSCP gels. It is added at concentrations of 1 to 10 per cent. It is unclear whether there is any benefit to heteroduplex detection provided by the addition of glycerol. The addition of a moderate amount of urea to heteroduplex analysis gels apparently improves resolution. This level of denaturant presumably is enough to loosen the helix, but without dissociating the DNA strands. The addition of urea may also help to sharpen the bands on the gel.

5.2.2 Fragment size

DNA fragments for heteroduplex analysis should be in the size range of 200–600 bp. Although the gel matrices can resolve fragments up to 1 kb, homoduplexes and heteroduplexes are difficult to resolve at sizes above 500 bp or so. Fragments for analysis are usually produced by PCR amplification. As mutants lying in the middle of DNA fragments are said to have a higher detection rate than those towards the ends of fragments, the detection rate can be maximized by designing PCR primers which amplify overlapping fragments or by amplifying long regions and digesting with restriction enzymes (the latter approach is obviously cheaper). Separate digestions with several four-cutter enzymes (those which recognize a 4 bp recognition site) should have much the same effect as overlapping PCR products (*Figure 5.32*).

5.3 Fragment visualization

5.3.1 Ethidium bromide staining

The simplest way to visualize the DNA is by ethidium bromide staining. Ethidium bromide stains DNA by intercalating into the helix. When stained, the DNA can be visualized and photographed by placing the gel on a UV transilluminator.

5.3.2 Silver staining

Silver staining of DNA in polyacrylamide gels is much less frequently used than ethidium bromide staining, as it is a longer multi-step procedure. It is also more expensive. It is, however, about five times

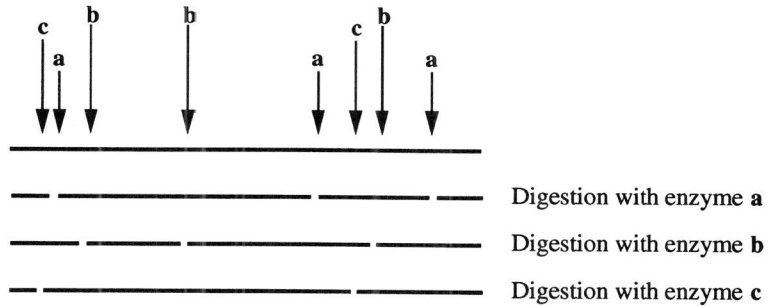

Fig 5.32

The creation of overlapping DNA fragments by restriction enzyme digestion.

Digestion with enzyme **a**

Digestion with enzyme **b**

Digestion with enzyme **c**

more sensitive than ethidium bromide staining. In heteroduplex analysis experiments, the heteroduplexes often stain at only about 25–30 per cent of the intensity of the homoduplex bands. The extra sensitivity provided by silver staining is therefore an attractive option. An additional advantage of silver staining is that, when dried, the gel can be kept, providing a permanent record without the need for photography.

5.3.3 Radiolabelling

In some examples of heteroduplex analysis, the users have chosen to radiolabel the DNA. It is unclear if and to what extent resolution is improved. Most heteroduplex analysis users choose not to label, as having chosen this method, they are willing to forego some sensitivity for the sake of simplicity.

5.3.4 Band behaviour on gels

◇ The smaller the DNA fragment size, the higher the gel concentration should be.

The homoduplex bands migrate according to their size, so the gel should be of a concentration appropriate to the fragment size. Size markers should be included. Heteroduplex bands, for reasons outlined earlier, usually migrate more slowly than the homoduplexes. When a mutation is a single base substitution, two bands will usually be present; one representing the two homoduplexes and one representing the two heteroduplexes (*Figure 5.33*, lane 2). Sometimes three bands are present—one representing the two homoduplexes and the other two bands representing the two heteroduplexes (*Figure 5.33*, lane 3). On rare occasions a fourth band may be present, which is due to separation of the two forms of the homoduplex molecule (*Figure 5.33*, lane 4). When the mutation is a small insertion or deletion, four bands are commonly observed.

The banding pattern produced is repeatable from gel to gel, and thus common polymorphisms and mutations can be identified from their banding pattern alone, without the need for further identification. Hence heteroduplex analysis is often also used for the scoring of known alleles.

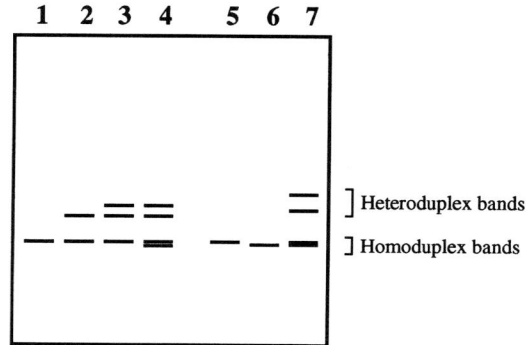

Fig 5.33

Different forms of DNA migration pattern encountered in heteroduplex analysis. Lanes 2–4, different point mutation patterns. Lane 6, homoduplex deletion. Lane 7, heteroduplex deletion. Lanes 1 and 5 are wild-type controls.

5.4   Generating the heteroduplex

When conducting mutation detection experiments by heteroduplex analysis, it is important to know whether any mutations are likely to be present in the heterozygous or homozygous state. Generally, the mutations responsible for dominant phenotypes are in the heterozygous state (i.e. on one allele only). The mutations responsible for recessive phenotypes may either be in the homozygous state (i.e. present on both alleles) or be compound heterozygotes (i.e. different mutations on each allele). In the case of recessive disorders, it is also useful to know which individuals are obligate carriers, and therefore heterozygous for the mutation.

In the case of samples known or expected to be heterozygous, heteroduplexes will form during PCR amplification. Unlike DGGE, where a sufficient level of heteroduplex is present at the end of the amplification, it is advisable with heteroduplex analysis to conduct an extra denaturation/annealing step to maximize the level of heteroduplex molecules. In the case of samples known or expected to be homozygous, an equal quantity of sample and wild-type PCR product should be mixed, heated, and cooled to generate the heteroduplex molecules.

In cases where you are unsure whether any mutation will be heterozygous or homozygous, two DNA samples should be analysed on the gel, one being the sample PCR product alone and the other being the mixed sample/wild-type PCR product.

5.5   Methodology

5.5.1 Reagents

The only reagents required for heteroduplex analysis are those required to make the gel and visualize the DNA. These are:

● acrylamide/bis (19:1 or 29:1) or MDE gel solution.
● ammonium persulphate
● TEMED

- glycerol (optional)
- urea (optional)
- TBE buffer
- loading dye
- ethidium bromide or silver stain

5.5.2 Making the gel

It is recommended that MDE gel solution is used in preference to acrylamide. The preparation of the gel plates and pouring of the gel should be performed as described in Section 1.2.1. It is preferable that a well-forming comb is used rather than a sharkstooth comb.

MDE gel solution is obtained as a 2× concentrate. Final concentrations of 0.6× to 1× should be used, depending on DNA fragment size. The inclusion of 15 per cent urea is recommended, as it may improve resolution. The buffer in the gel mix is provided by 0.6× TBE.

5.5.3 Running the gel

◇ Loading dye is 40 per cent sucrose, 0.05 per cent bromophenol blue, 0.05 per cent xylene cyanol.

Genomic DNA or cDNA should be PCR-amplified to give a clean product. To 5 μl of heteroduplexed PCR product, you should add 1 μl of loading dye. Prior to loading, the wells should be flushed out with a Pasteur pipette and all of the sample loaded. It is recommended that a positive control sample (a known mutation) is always included.

◇ On a 1× MDE gel, the bromophenol blue runs at about 70 bp and the xylene cyanol at about 230 bp.

The gel should be run in 0.6× TBE at 800 V. For DNA fragments over 200 bp, it will probably be necessary to run the gel overnight. The DNA fragments should be run at least two-thirds of the way down the gel. When the dye has migrated to a satisfactory point, the power supply should be turned off and the gel apparatus should be dismantled.

5.5.4 Staining the gel

After separating the gel plates, the gel, still attached to one plate, should be ethidium bromide stained and visualized as described in Section 4.2.5.

If the gel is to be silver stained, the gel should be placed in 10 per cent ethanol for 5 minutes with gentle agitation. The ethanol solution should be carefully discarded and then the gel be allowed to oxidize in 1 per cent nitric acid for 3 minutes. The nitric acid should be discarded and the gel briefly rinsed with water. The gel should then be soaked on 0.012 M silver nitrate for 20 minutes. Following this, you should decant the silver solution and very briefly rinse the gel in water. The gel should then be reduced in 0.28 M sodium bicarbonate (anhydrous)/0.019 per cent formalin. The solution will turn brown, and when it does so, should be replaced. Several solution replacements may be necessary. When the image has developed, the process should be stopped with 10% glacial acetic acid. After 5 minutes the solution should be discarded and replaced with water for a further 5 minutes. Any excess silver deposits on the gel can be removed with cotton wool balls. The developed gel can be stored following air-drying.

5.6   The result

An ideal result is illustrated in *Figure 5.34*. On a 40 cm-long gel, homoduplex and heteroduplex bands should separate by at least 3 mm. Mutations are characterized by cloned or direct sequencing of PCR products.

Fig 5.34

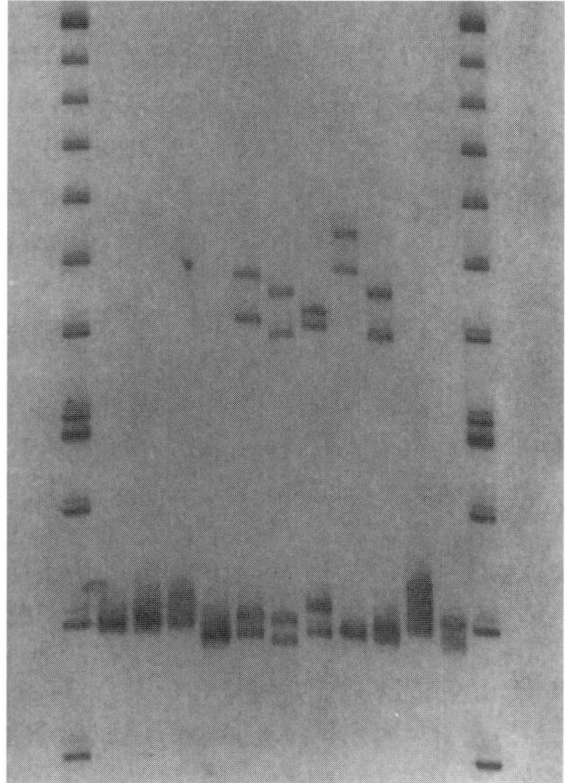

Heteroduplex analysis detection of mutations in the human CFTR gene on an MDE gel. (Reproduced with permission from Highsmith, 1993; American Association for Clinical Chemistry, Inc.).

5.6.1 Problems

The loading of too little or too much PCR product can cause the invisibility of heteroduplex bands or the merging of heteroduplex and homoduplex bands, respectively. Excessive ethidium staining can also mask the visibility of heteroduplex bands. Extended periods of destaining can cure this. Above all, as with the other mutation detection methods, the yield of the PCR product needs to be good and background misamplification must be low.

6.   The protein truncation test

Certain genes when mutated to produce a particular phenotype show a preponderance of 'severe' mutations. These may be frameshifts or more simply point mutations which cause premature protein termination (nonsense mutations).

The protein truncation test (PTT), eliminates the need to scan each base of the DNA by simply testing the size of the protein product, which is synthesized *in vitro* from a PCR product.

The technique is relatively simple and provides a rapid way to screen large regions for 'severe' mutations. As the method is unable to detect 'subtle' mutations, i.e. missense mutations, the results cannot be confused by the presence of polymorphic silent variants.

6.1 The theory

Some genes appear to be very tolerant of missense mutations and require more severe mutations such as nonsense mutations, splicing defects, or deletions to cause an abnormal phenotype. The tumour-suppressor genes, *APC*, *NF1*, and *BRCA1 & 2*, as well as the Duchenne muscular dystrophy gene, *DMD*, fall into this category. For all of these genes and others like them, mutations would be most readily detected by analysing the protein products rather than the genes. Obtaining the relevant proteins from the patients would be invasive, painful, and difficult. The proteins can however be synthesized (in pieces if necessary) *in vitro* from PCR templates.

For the PCR amplification, the forward (or left-hand) primer, has attached a T7 RNA polymerase promoter and a eukaryotic translation initiation sequence *Figure 5.35*. Protein products are synthesized in a coupled transcription/translation reaction and analysed by polyacrylamide gel electrophoresis (Roest *et al.* 1993). Truncated proteins are

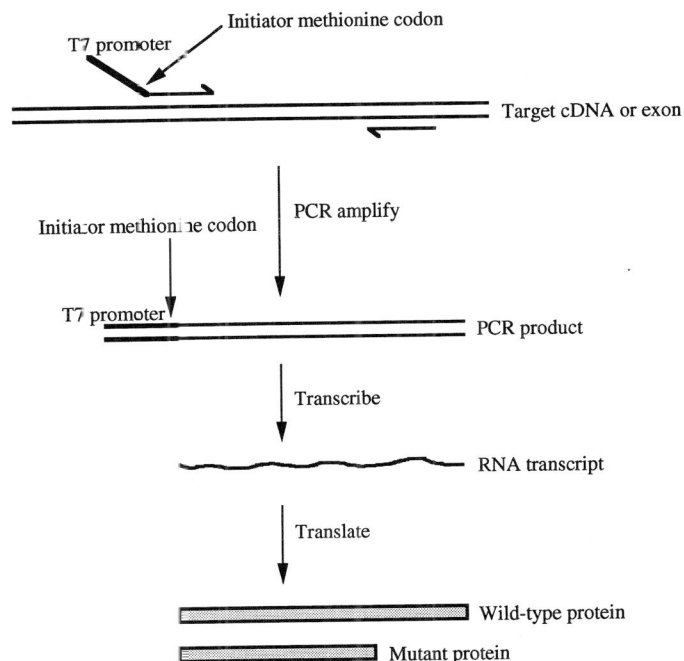

Fig 5.35

The protein truncation test methodology.

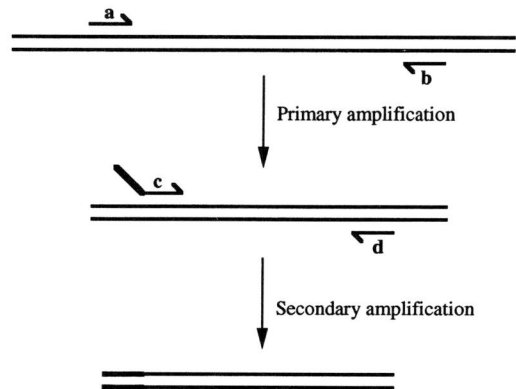

Fig 5.36

Nested PCR amplification. A second pair of primers (c and d) are 'nested', i.e. are internal to the primary primers (a and b).

readily discriminated from the wild-type proteins (Heim *et al.* 1994; van der Luijt *et al.* 1994;) and their size directly pinponts the position of the mutation.

6.1.1 PCR amplification

In order to obtain a good yield of clean PCR product, it may be necessary to conduct two rounds of amplification, the second using a nested pair of primers. Two rounds of amplification will definitely be required if you are doing RT-PCR on trace levels of RNA (see Chapter 2, Section 1.2). The forward nested primer (primer c in *Figure 5.36*) has attached to its 5′ end the T7 RNA polymerase promoter sequence and the 'ATG' eukaryotic translation initiation signal. The sequence of this primer tail is 5′-GGATCCTAATACGACTCAC-TATAGGAACAGACCACCATG........-3′. The gene-specific 3′ part of the primer should be designed such that the ATG initiator codon is in-frame with the coding sequence of the gene.

A great advantage of the PTT method is that it can scan large regions of a gene rather than, for example SSCP, which can only

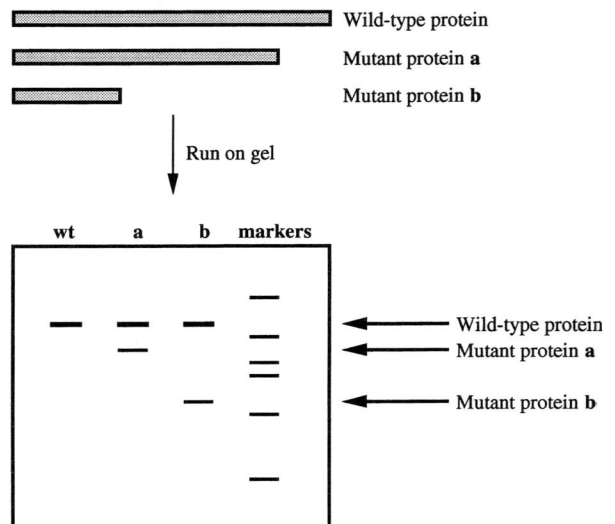

Fig 5.37

Proteins derived from two heterozygous mutant genes.

examine very small fragments. The DNA source of choice is therefore cDNA, as large contiguous fragments of coding sequence can be amplified on a single fragment. Exons of the genomic DNA will often be too small for PTT analysis, with only unusually large exons being suitable. If conducting PTT on cDNA, you should always bear in mind the possibility that mutant alleles can be missed due to RNA instability (see Chapter 2, Section 1.4).

6.1.2 *In vitro* transcription and translation

The transcription and translation is conducted using a commercially available kit in which both reactions occur simultaneously in the same tube. The system uses T7 RNA polymerase to synthesize RNA from the PCR product and rabbit reticulocyte lysate to synthesize protein from the RNA. A cocktail of amino acids is included but lacking in methionine. The protein produced is radiolabelled by the inclusion of ^{35}S-methionine.

6.1.3 Electrophoresis

Denatured protein samples are run on small SDS-polyacrylamide gels. When the run is complete, the gel is fixed and then soaked in an organic scintillant to increase band detection sensitivity. The scintillant converts the emitted energy of the isotope into light and thus increases the proportion of energy which is detectable by X-ray film at –70 °C.

6.1.4 Mutation detection and mapping

Mutations are detected by the presence of a protein band of a smaller size than the wild-type. The size of the protein also defines the position of the mutation, as the protein terminates at the position of the stop codon. If, for example the wild-type protein produced is 55 kDa and the mutant protein is 30 kDa, it is possible to work out the approximate position of termination in the gene. As a rough guide, nine amino acids approximates to 1 kDa. The 55 kDa protein is approximately 495 amino acids in length and the mutant 30 kDa protein, 270 amino acids. 495 amino acids corresponds to 1485 bp and 270 amino acids to 810 bp. The position of termination is thereforeapproximately 810 bp 3' to the ATG codon on the forward PCR primer.

It should be noted, however, that if the truncated protein has been caused not by a nonsense mutation, but by a frameshift mutation, the insertion, deletion, or splicing defect responsible, may lie anywhere 5' to the translation termination site.

As mutations at positions corresponding to the ends of the fragment might result in a protein band too small to see or too large to be resolved from the wild-type, it is advisable that PCR products should overlap (*Figure 5.38*).

6.2 Methodology

If the fragment to be amplified is over 1 kb, it may be necessary to use a polymerase and buffer system appropriate for long-distance PCR

Fig 5.38
___

The use of overlapping PCR fragments in PTT. The mutant protein derived from fragment (a) would be indistinguishable from the wild-type. Whereas the mutant protein derived from fragment (b) would be readily distinguished.

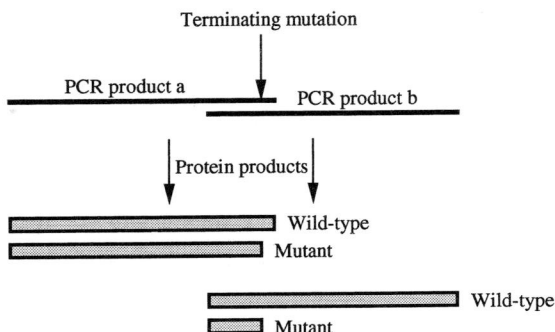

amplification. If a secondary amplification is required, 1 μl of the PCR product should be diluted in 100 μl of water. The diluted primary product (1 μl) should then be re-amplified with internally nested primers. The forward primer (or forward nested primer) should have the transcription/translation tail attached.

6.2.1 Coupled transcription/translation

The following mix is enough for 10 reactions:

◇ All of the transcription/translation components except the labelled methionine and RNase inhibitor are supplied in the TNT™ coupled reticulocyte lysate kit from Promega.

	μl
rabbit reticulocyte lysate	62.5
reaction buffer	5
T7 RNA polymerase	2.5
amino acid mix minus methionine	2.5
35S-methionine	5
RNase inhibitor (40 u/ul)	2.5
TE	20
	100

◇ The DTT should only be added to the sample buffer, just before use.

The mix (10 μl) should be added to 2 μl of the PCR product. The reaction tube should be incubated at 30 °C for 1–2 hours. The reaction mix (3 μl) should then be added to 12 μl of the sample buffer (20 per cent glycerol, 2 per cent SDS, 0.0025 per cent bromophenol blue, 0.125 M Tris pH 6.8, and 100 mM DTT).

6.2.2 Pouring and running the gel

The gels consist in the main part of the separating gel, which is overlayed with a low percentage acrylamide stacking gel. The purpose of the stacking gel is to concentrate the sample, which is loaded in a relatively large volume, resulting in better band resolution. The acrylamide concentration of the separating gel will depend upon the size of the proteins to be loaded.

◇ The separating gel is 8–17 per cent acrylamide (2.6%C), 0.375 M Tris pH 8.8, 0.1 per cent SDS. The stacking gel is 4 per cent acrylamide, 0.125 M Tris pH 6.8, 0.1 per cent SDS.

A small gel apparatus such as Bio-Rad Protean II or Hoeffer Mighty Small II is ideal.

◇ Ammonium persulphate (10 per cent) and TEMED should be added to the separating gel mix at 0.5 and 0.05 per cent, respectively and to the stacking gel mix at 0.5 and 0.1 per cent, respectively.

Before pouring the gel, the comb should be inserted into the gel assembly and a line 1 cm below the teeth of the comb should be drawn on the plate with a marker pen. This line marks the level to which the

◇ A rough guide to appropriate
acrylamide concentrations.
  40–200 kDa, 8 per cent
  21–100 kDa, 10 per cent
  10–40 kDa, 12 per cent

◇ The running buffer is Tris/glycine
buffer (0.025 M Tris base, 0.2 M glycine,
0.1 per cent SDS)

◇ Fixing solution is 10 per cent acetic
acid, 20 per cent methanol.

separating gel should be poured. The gel should be poured down the edge of a spacer and up to the line. The gel should be immediately overlayed with water-saturated butan-2-ol using a Pasteur pipette. This has the effect of creating a flat surface and excluding oxygen. After the gel has polymerized (this should take 45–60 minutes), you should rinse off the overlay solution with water and dry the area above the separating gel with blotting paper.

The comb should be placed in the assembly and tilted slightly so that air bubbles will not get trapped under the teeth. The stacking gel mix should be poured into the gel assembly down the spacer at the upturned side of the comb. When all the teeth are covered, the comb should be put properly in place, and the gel mix poured to the top. When the gel has set, the comb should be removed and the wells flushed out with water. The gel apparatus should then be fully assembled.

The samples should be boiled for 2 minutes to fully denature the protein. The whole volume is then loaded onto the gel. A size marker track should be included, e.g. Amersham's ^{14}C rainbow markers. The gel should be run at 20 mA until the bromophenol blue dye reaches the bottom. The unincorporated label migrates with the dye. It is a matter of preference whether this is allowed to run into the buffer or is kept at the bottom of the gel. The gel takes approximately 2 hours to run. The gel apparatus should then be disassembled and the gel soaked in fixing solution for 30 minutes.

The gel is next soaked in an organic scintillant, such as Amersham's Amplify™, for 30 minutes. The gel is then placed onto 3MM paper, covered in Saran Wrap and dried. The Saran Wrap is then removed and the gel exposed to X-ray film at –70 °C overnight.

6.3 The result

An ideal result is shown in *Figure 5.39*. It is essential to include several samples from unrelated individuals on the gel, so that abnormally small protein products can be differentiated from degradation products.

6.3.1 Problems

The TNT™ kit comes complete with a positive control, so that a problem with the transcription/translation components is easy to eliminate.

The most likely source of problems is in the PCR step. Getting good yields from RT-PCRs is not always easy, especially if amplifying from trace levels of RNA derived from lymphocytes.

7. Fluorescence-based DNA sequencing

Modern fluorescent-based direct cycle-sequencing is rapid enough to be considered a viable approach to mutation screening. The sequence produced has to be of top quality in order to detect mutations in the

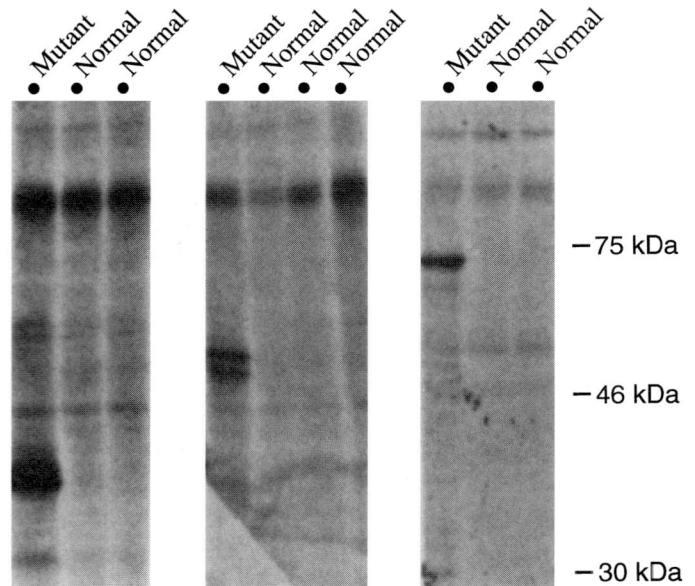

Fig 5.39

PTT mutation detection in the human APC gene in three individuals. (Reproduced with permission from Dobbie *et al.* 1996; BMJ Publishing Group).

◇ Advantages
● Nonisotopic
● Characterizes mutation

◇ Disadvantages
● Labour intensive
● Multi-step procedure
● Multiple sequence reads required to get 100 per cent accuracy
● Relatively slow

heterozygous state. A single unreadable base in a sequence trace necessitates resequencing. It is this requirement for multiple sequence runs that discourages many from adopting this approach to mutation detection. The speed of improvements to DNA sequencing technology however, may mean that sequencing might one day become the common method of mutation detection.

7.1 Alternative methods of DNA sequencing

The presence of a chapter on the identification of mutations by DNA sequencing demonstrates both the failure of the various mutation detection methods to live up to our hopes and needs, and also the progress made in speeding up DNA sequencing.

There are dozens of alternative strategies and varieties of chain termination (or dideoxy) DNA sequencing. Only an outline of the most common approaches is presented here. There are several choices to be made: do you do cloned or direct, isotopic or fluorescent, cycle or noncycle sequencing; should the label be incorporated or fixed to the primer, and which enzyme is best? The decision made depends on what is the easiest and fastest approach for the project concerned. It should be noted that the ability to obtain good quality sequence data, depends only a little on your sequencing skills, but mostly on the quality of the DNA template to be sequenced.

7.1.1 Cloned versus direct sequencing

Until the advent of PCR, virtually all examples of DNA sequencing were performed on cloned templates. This was usually in the form of small fragments subcloned into the single-stranded bacteriophage

M13 or into pBR322-based double-stranded plasmids. The purpose of cloning the fragment is to produce a large quantity of pure and clean DNA: a prerequisite for sequencing. This may be considered to be 'biological' or *in vivo* amplification. The introduction of PCR enabled large quantities of defined DNA fragments to be produced without the need for cloning and subcloning, i.e. *in vitro* amplification.

Until methods for the sequencing of PCR products directly were developed, they were cloned (generally into plasmids) prior to sequencing. There were several problems associated with this approach. The first was a problem of cloning itself, due to the generation of 3' adenosine overhangs on the PCR product by *Taq* DNA polymerase, cloning into blunt-ended vectors was difficult. The second problem was due to the appearance of 'unreal' mutations. This is due to the infidelity of *Taq* DNA polymerase, allowing bases to be misincorporated at a frequency of about 1/2000 base pairs. Thus, when a molecule in a PCR product carrying an error is cloned, all plasmids derived from this molecule will carry the error, so when sequenced, the error will appear as a mutation in the sequence. Thus, when any variant is identified, it requires verification by the sequencing of multiple clones. Heterozygosity poses another problem. A high proportion of single-gene diseases in humans, and altered phenotypes in other diploid organisms, are the result of a mutation in only one allele or different mutations on the two alleles (compound heterozygosity). When sequencing cloned DNA, it is not (usually) possible to know which allele you are sequencing. To be sure that you end up sequencing both alleles, you need to sequence at least five clones as there is a risk that you might repeatedly sequence the same allele (see Section 3.2.2).

The ability to sequence the PCR product DNA directly, has largely overcome these problems. The cloning steps are removed completely, and *Taq*-induced errors are not seen, as a population of molecules is being sequenced together rather than (effectively) a single molecule. An error in one molecule is not seen, as it is masked by the correct base at that site in the vast majority of molecules.

The problem of heterozygosity has unfortunately, not gone away, but changed. In direct sequencing, at the position of a heterozygous mutation, two different bases will be detected at the same point. In order to tell that there really are two bases at this point, rather than this just being a bit of poor quality sequence (known as a 'stop'), the overall quality of the sequencing needs to be very good. Another draw-back of direct sequencing is that the PCR product usually needs to be purified to eliminate unincorporated primers and nucleotides as well as nonspecific amplification products. There is, therefore, no substitute for fragment isolation from gels.

7.1.2 Isotopic versus fluorescent sequencing

This may also be considered to be manual versus automated sequencing. Automated (or semi-automated) sequencing is so-called because the fluorescent labels in the gels are read and interpreted by machine.

Until the 1990s virtually all DNA sequencing was isotopic, the usual label being ^{35}S. The method of choice is now very definitely fluorescent sequencing but, due to the high cost of the equipment required, many laboratories do not have the choice. Fluorescent sequencing has many advantages:

(1) it is nonisotopic;

(2) more bases can be read from a single reaction;

(3) all four reactions are loaded into a single gel track;

(4) the gel is read by laser and the data is directly downloaded into a computer.

Points 2, 3, and 4 underline the overall speed of fluorescence-based sequencing as compared with isotopic sequencing.

One attraction of conventional sequencing in mutation detection is that the data can be directly and clearly visualized by inspection of the resulting autradiograph. If each of the four termination reactions are run separately, i.e. the A-tracks on one gel, the T-tracks on another, etc., mutations can be easily spotted at a glance, without needing to read the sequence. This approach is best suited to haploid genomes, where the gain of a band on one gel, is accompanied by the loss of a band from another.

This same approach can be applied with a little effort to automated sequencers and can even allow several samples to be loaded in a single track (Hattori *et al.* 1993).

7.1.3 Dye-primers versus dye-terminators

It is usual in isotopic sequencing that one nucleotide is labelled and is incorporated during the extension step of the reaction. Only rarely is the label attached to the primer. In fluorescent-based sequencing, however, both extension incorporation using dye-labelled dideoxy-nucleotides (dye-terminators) and dye-labelled primers are commonly used as the labelling method. When dye-terminators are used, each ddNTP is labelled with a different coloured dye and a single reaction is conducted incorporating all four ddNTPs simultaneously. With dye-primers, four differently labelled versions of the primer are required. Each label is not specific to a particular ddNTP, so each dye has to be used with a single ddNTP in separate reactions.

Fluorescent sequencing requires much more DNA on the gel than isotopic sequencing. A rapid alternative to starting with large quantities of clean template DNA is to perform 'cycle sequencing' on the PCR product. This is a process in which multiple rounds of sequencing reactions are performed, giving a linear amplification of sequence product. *Taq* DNA polymerase, which is used for the reaction is poor at incorporating dideoxynucleotides, resulting in variable signal intensities at each base position. It is considerably easier, when attempting to identify heterozygous mutations, to spot a double peak (i.e. two different bases at the same position) when peak heights (i.e. signal intensities) are even. This is more readily achieved when using dye-

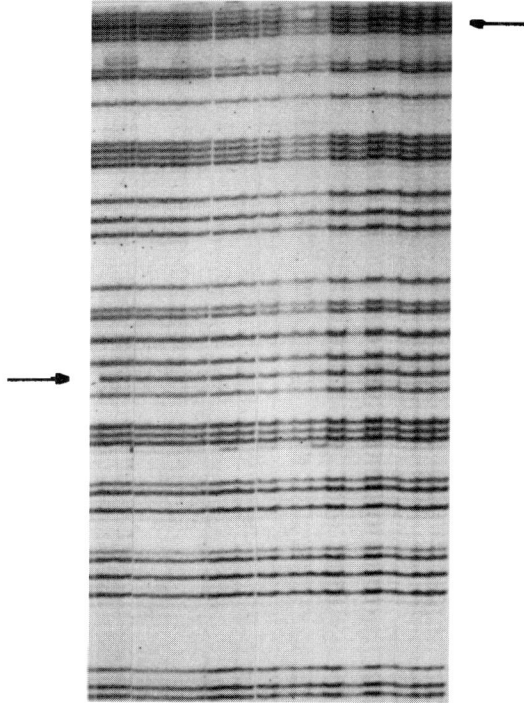

Fig 5.40

Mutation detection in a haploid gene by single-track sequencing. In this 'A-track' gel, two samples are shown to be mutant by the loss of a band (indicated by arrows). In the mutant samples an extra band appears on one of the other three single track gels.

primers than dye-terminators, as dye-terminators are even more difficult to incorporate than unlabelled terminators.

Heterozygous mutation sequencing using dye-terminators requires the peak trace to satisfy several criteria before a mutation can be called: namely a greatly diminished signal, the presence of a newly emerging base underneath the peak, and also a significant change in peak height in the immediately 3′ adjacent base as compared with a known wild-type sequence trace (Kwok *et al.* 1994). The use of dye-primers is therefore preferable, but the cost of dye-primers is very high. A cost-reducing approach which has made mutation screening feasible, has been to tail the PCR primers with the complementary sequence of the 'standard' M13 forward and reverse primers. This enables all of the sequencing to be performed with a single pair of labelled (M13) primers, provided that the PCR product is small enough for the entire sequence to be obtained from a single read, i.e. <400 bp. The extra cost of needing long (35–40 bp) PCR primers is vastly offset by the savings on labelled primers.

Fig 5.41

In poor quality sequence output, peak heights are uneven, background is high. Heterozygous point mutations are indistinguishable from unreadable bases. Good quality sequence output is characterized by even peak heights and low background. Heterozygous mutations are characterized by a double peak of about half the usual height.

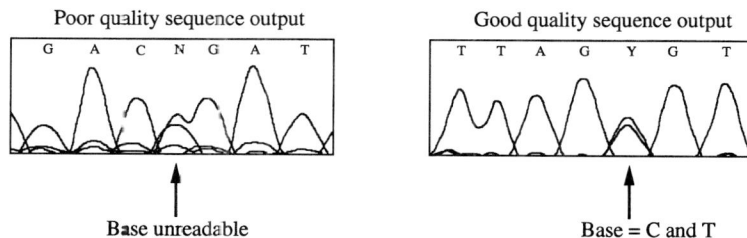

Poor quality sequence output

G A C N G A T

Base unreadable

Good quality sequence output

T T A G Y G T

Base = C and T

An additional bonus of dye-primer sequencing is that less template DNA is required.

7.1.4 Conventional sequencing versus cycle sequencing

The production of good quality sequence data may be produced by the cycle sequencing of a double-stranded PCR product using a thermostable DNA polymerase (see above), or by the purification of large quantities of single-stranded DNA for conventional sequencing. The use of Streptavidin-coated plastics and magnetic beads has made a big impact on this latter approach. The PCR product, after purification, maybe attached to the solid support via a Biotin group on one primer. The DNA strands are separated by alkaline denaturation and the unattached strand removed. Either strand may be used as a template for sequencing, however, sequencing of the bound strand is easier. The polymerase of choice is 'Sequenase™' (T7 DNA polymerase, Amersham), as it does not need to be thermostable and produces excellent, even signal intensities. However, unlike cycle sequencing, Sequenase conventional sequencing requires a large amount of template DNA.

The introduction of genetically modified versions of *Thermus aquaticus* (*Taq*) DNA polymerase specifically tailored for DNA sequencing such as AmpliTaq FS™ (Perkin Elmer) and Thermo-Sequenase™ (Amersham) give much more even signal intensities than ordinary *Taq* DNA polymerase, and so combine the advantages of Sequenase sequencing with those of cycle sequencing.

7.2 DNA sequencing as a mutation detection technique

Manual (or isotopic) sequencing is slow and very labour intensive. It is only a practical approach to mutation screening when the gene or region to be sequenced is very small. In the more usual instance of screening regions of several kilobases in size, the only viable sequencing approach is that of automated (or fluorescent) sequencing. Whether you are screening for homozygous or heterozygous mutations, the quality of the sequence needs to be very good. A 400-base read with just one unreadable base will have to be repeated. It is this phenomenon of requiring multiple reads to obtain the full sequence which makes DNA sequencing a rarely used approach to mutation detection.

A further disadvantage of mutation detection by direct sequencing is that deletions and insertions are not always obvious. Beyond the point of the deletion or insertion on the sequence trace, the sequence becomes unreadable, as two different sequences are being overlaid. It is not always clear whether this is due to the presence of a deletion or insertion, or simply due to the breakdown of the sequencing. In all cases of a sequence abnormality, the opposite strand must be sequenced to confirm it as a mutation.

It should be noted that a mutation detected by any means other than DNA sequencing, has to be sequenced to be characterized. The advantage of the sequencing approach is that it combines the detection and characterization steps.

Given the rate at which sequencing technology is developing, unless a new and ideal scanning technique is soon devised (see next chapter), sequencing will become a more and more widely adopted method of mutation detection.

8.   Choosing a method

The main choices to be made in deciding which technique to use to identify unknown mutations are:

- Do I need to detect 100 per cent of all mutations?
- What level of experimental difficulty am I prepared to tolerate?
- Is any particular technique well or poorly suited to the question I am asking?

If you have only one or a very small number of samples to test, it is essential that the technique you use is not going to miss the mutation. In this case you will require a technique with high efficacy such as DGGE, CCM, or DNA sequencing. If, however, you have the alternative scenario of equating a candidate disease gene with a particular disease and you have many patient samples, you may wish to choose a technique which will give you your answer rapidly, such as SSCP or PTT, and be happy not to detect all mutations.

The ideal mutation detection technique would be both simple to perform and have a 100 per cent detection rate. Until that technique is invented, you may be forced through lack of experience or equipment, to perform a simple method. The simplest technique, in terms of equipment required, user experience, and detection ease is heteroduplex analysis. This technique though has probably the lowest detection rate of all. SSCP is the next simplest, but is best performed in a cold room and using radioactive or fluorescent detection. If simplicity is the major driving force, the use of a fan to cool a room-temperature run gel and silver staining of the gel, can allow SSCP to be very simple, but at the cost of the detection rate.

Of the techniques with very high detection rates (i.e. DGGE, CCM, DNA sequencing), DGGE is probably the simplest; the main difficulty being in experimental design rather than in executing the experiment. DNA sequencing is probably the most difficult technique for detecting mutations. This is because, although not technically difficult to carry out, every last base pair must be called with a high degree of confidence. In reality this means 90 per cent of the data is collected in 10 per cent of the time and the final 10 per cent of the data takes 90 per cent of the time to collect. If you have only a very small number of samples to test, DNA sequencing is an attractive technique. But for larger sample numbers, it is only a viable proposition if you are happy with your sequencing ability and cannot be bothered to set up any new techniques.

A major factor in choosing a technique, is DNA fragment size. If your gene of interest is large, to minimize the amount of work involved you might only be prepared to consider a method capable of handling large fragments, such as CCM.

Cost is not such a major concern with the mutation scanning techniques as the diagnostic techniques, as all of the more effective methods are quite expensive. Of the most expensive are DNA sequencing, which can get costly if many runs are required to get all the bases read with confidence, and the protein truncation test, due to the high cost of *in-vitro* translation reagents.

Safety concerns are a major factor in choosing a technique. The use of radioisotopes may not be desirable or not possible if your laboratory does not have a usage licence. The conversion of these techniques to fluorescent detection is beginning to reduce the dependence on radioisotopes, but requires very expensive equipment, which is beyond the limits of most laboratory budgets. The requirement of toxic chemicals, such as formamide in DGGE and osmium tetroxide and piperidine in CCM, may be a discouraging factor. The availability of a fume hood is essential in these techniques. It is especially important that female laboratory workers in the reproductive age must not be exposed to teratogenic chemicals as, if pregnant some risk is put on the normal development of the fetus.

All of the above factors will influence which technique you choose to use. Although some of these techniques are easier to perform than others, it should be stressed that none of these techniques are easy. All require gel electrophoresis and in some cases use very sophisticated gel systems. Some skill will be required to get the techniques working to their best and some experience may be required in differentiating mutations from background 'noise'. As with any reasonably difficult technique, it is essential that you follow the protocol exactly (in the initial experiments at least). Problems should be expected. In the event of problems, it is advantageous to be able to discuss them with someone who has experience of the method. If that is not possible, don't be afraid of writing to or calling someone who has extensive experience of the method, best of all – the inventor. The prospect of problems should not put you off. Mutation detection is one of the most important areas of biology today, with much of our knowledge of biological function being based upon mutation data. The rewards of mutation detection and characterization can be very great. Good hunting!

Further reading

Brown, T. A. (1994). *DNA sequencing: the basics*, IRL press at Oxford University Press, Oxford. The complete backgound to DNA sequencing.

Gardner, R. J., Bobrow, M., and Roberts, R. G. (1995). The identification of point mutations in Duchenne muscular dystrophy patients by using reverse-transcription PCR and the protein truncation test. *American Journal of Human Genetics*, **57**, 311. A very clear article on the theory, methodology, and uses of the protein truncation test.

Hawkins, J. R. and Dalgleish, R. (1991). The detection and mapping of point mutations by RNase A cleavage. In *Methods in molecular biology*, Vol 9 (ed. C. G. Mathew), pp. 111–21, Humana Press, Clifton, NJ. Provides experimental details on riboprobe labelling and hybridization of the probe to RNA, rather than DNA.

Hovig, E., Smith-Sorensen, B., and Borresen, A.-L. (1994). Detection of mutations by denaturing gradient gel electrophoresis. In *Current protocols in human genetics* (ed. N. C. Dracopoli, J. L. Haines, B. R. Korf, D. T. Moir, C. C. Morton, C. E. Seidman, J. G. Seidman, and D. R. Smith), pp. 7.5.1–7.5.12, Wiley, New York. A thorough description of all forms of denaturing gradient gels.

Howells, D. W., Forrest, S. M. Dahl, H.-H. M., and Cotton, R. G. H. (1990). Insertion of an extra codon for threonine is a cause of dihydropterine reductase deficiency. *American Journal of Human Genetics*, **47**, 279. This paper contains a good account of the experimental details of CCM analysis.

Myers, R. M., Maniatis, T., and Lerman, L. S. (1987). Detection and localization of single base changes by denaturing gradient gel electrophoresis. *Methods in Enzymology*, **155**, 501. A dated, but thorough, account of denaturing gradient gels.

Saleeba, J. and Cotton, R. G. H. (1993). Chemical cleavage of mismatch to detect mutations. *Methods in Enzymology*, **217**, 287. Provides a protocol for the CCM method and gives details of the best source of the main chemicals required.

Watkins, H. C. (1994). Cleavage of RNA–DNA hybrids at mutation sites using RNase A. In *Current protocols in human genetics*, (ed. N. C. Dracopoli, J. L. Haines, B. R. Korf, D. T. Moir, C. C. Morton, C. E. Seidman, J. G. Seidman, and D. R. Smith), pp. 7.2.1–7.2.7, Wiley, New York. A thorough account of the RNase cleavage methodology.

Wood, N., Tyfield, L., and Bidwell, J. (1993). Rapid classification of phenylketonuria genotypes by analysis of heteroduplexes generated by PCR-amplifiable synthetic DNA. *Human Mutation*, **2**, 131. Describes the use of a 'universal heteroduplex generator' to aid the detection of known mutations by heteroduplex analysis.

Comparative PCR sequencing: A guide to sequencing-based mutation detection, Perkin Elmer (Stock No. 770901-001). A guide to mutation detection by fluorescence-based sequencing.

References

Brown, T. A. (1994). *DNA sequencing: the basics*, IRL Press at Oxford University Press, Oxford.

Costes, B., Girodon, E. Ghanem, N., Chassignol, M., Thuong, N. T., Dupret, D., *et al.* (1993). Psoralen-modified oligonucleotide primers improve detection of mutations by denaturing gradient gel electrophoresis and provide an alternative to GC-clamping. *Human Molecular Genetics*, **2**, 393.

Cotton, R. G. H., Rodrigues, N. R., and Campbell, R. D. (1988). Reactivity of cytosine and thymidine in single-base-pair mismatches with hydroxylamine and osmium tetroxide and its application to the study of mutations. *Proceedings of the National Academy of Sciences, USA*, **85**, 4397.

Dobbie, Z., Spycher, M., Mary, J.-L., Häner, M., Guldenschuh, I., Hürliman, R., *et al.* (1996). Correlation between the development of extracolonic manifestations in FAP patients and mutations beyond codon 1403 in the APC gene. *Journal of Medical Genetics*, **33**, 274.

Fischer, S. G. and Lerman, L. S. (1983). DNA fragments differing by single base pair substitutions are separated in denaturing gradient gels: correspondence with melting theory. *Proceedings of the National Academy of Sciences, USA*, **80**, 1579.

Fodde, R. and Losekoot, M. (1994). Mutation detection by denaturing gradient gel electrophoresis. *Human Mutation*, **3**, 83.

Forrest, S. M. Dahl, H. H., Howells, D. W., Dianzani, I., and Cotton, R. G. H. (1991). Mutation detection in phenylketonuria by using chemical cleavage of mismatch: importance of using probes from both normal and patient samples. *American Journal of Human Genetics*, **49**, 175.

Glavač, D. and Dean, M. (1993). Optimization of the single-strand conformation poly-morphism (SSCP) technique for detection of point mutations. *Human Mutation*, **2**, 404.

Grompe, M., Muzny, D. M., and Caskey, C. T. (1989). Scanning detection of mutations in human ornithine transcarbamoylase by chemical mismatch cleavage. *Proceedings of the National Academy of Sciences, USA*, **86**, 5888.

Haris, I. I., Green, P. M., Bentley, D. R., and Giannelli, F. (1994). Mutation detection by fluorescent chemical cleavage: application to hemophilia B. *PCR Methods and Applications*, **3**, 268.

Hattori, M., Shibata, A., Yoshioka, K. and Sakaki, Y. (1993). Orphan peak analysis: a novel method for detection of point mutations using an automated fluorescence DNA sequencer. *Genomics*, **15**, 415.

Hawcroft, D. M. (1996). *Electrophoresis: the basics*, IRL Press at Oxford University Press, Oxford.

Heim, R. A., Silverman, L. M., Farber, R. A., Kam-Morgan L. N. W., and Lace, M. C. (1994). Screening for truncated NF1 proteins. *Nature Genetics*, **8**, 218.

Highsmith, W. E. (1993). Carrier screening for cystic fibrosis. *Clinical Chemistry*, **39**, 706.

Hovig, E., Smith-Sørensen, B., Brøgger, A. and Børresen, A.-L. (1991). Constant denat-urant gel electrophoresis, a modification of denaturing gradient gel electrophoresis, in mutation detection. *Mutation Research*, **262**, 63.

Iwahana, H., Yoshimoto, K., Mizusawa, N., Kudo, E., and Itakura, M. (1994). Multiple fluorescence-based PCR-SSCP analysis. *BioTechniques*, **16**, 296.

Keen, J., Lester, D., Inglehearn, C., Curtis, A., and Bhattacharya, S. (1991). Rapid detection of single base mismatches as heteroduplexes on Hydrolink gels. *Trends in Genetics*, **7**, 5.

Kwok P.-Y., Carlson, C., Yager, T. D., Ankener, W., and Nickerson, D. A. (1994). Comparative analysis of human DNA variations by fluoresence-based sequencing of PCR products. *Genomics*, **23**, 138.

Lerman, L. S. and Silverstein, K. (1987). Computational simulation of DNA melting and its application to denaturing gradient gel electrophoresis. *Methods in Enzymology*, **155**, 483.

Miyoshi, Y., Ando, H., Nagase, H., Nishisho, I., Horii, A., Miki, Y., *et al.* (1992). Germ-line mutations of the *APC* gene in 53 familial adenomatous polyposis patients. *Proceedings of the National Academy of Sciences, USA*, **89**, 4451.

Myers, R. M., Larin, Z., and Maniatis, T. (1985). Detection of single base substitutions by ribonuclease cleavage at mismatches in RNA:DNA duplexes. *Science*, **230**, 1242.

Nagamine, C. M., Chan, K., and Lau, Y.-F. C. (1989). A PCR artifact: generation of heteroduplexes. *American Journal of Human Genetics*, **45**, 337.

Orita, M., Iwahana, H., Kanazawa, H., Hayashi, K., and Sekiya, T. (1989*a*). Detection of polymorphisms of human DNA by gel electrophoresis as single-strand conforma-tion polymorphisms. *Proceedings of the National Academy of Sciences, USA*, **86**, 2766.

Orita, M., Suzuki, Y., Sekiya, T., and Hayashi, K. (1989*b*). Rapid and sensitive detec-tion of point mutations and DNA poymorphisms using the polymerase chain reaction. *Genomics*, **5**, 874.

Ravnik-Glavač, M., Glavač, D., Chernick, M., di Sant'Agnese, P., and Dean, M. (1994*a*). Screening for CF mutations in adult cystic fibrosis patients with a directed and optimized SSCP strategy. *Human Mutation*, **3**, 231.

Ravnik-Glavač, M., Glavač, D. and Dean, M. (1994*b*). Sensitivity of single-strand con-formation polymorphism and heteroduplex method for mutation detection in the cystic fibrosis gene. *Human Molecular Genetics*, **3**, 801.

Roest, P. A. M., Roberts, R. G., Sugino, S., van Ommen, G. J. B. and den Dunnen, J. T. (1993). Protein truncation test (PTT) for rapid detection of translation-terminating mutations. *Human Molecular Genetics*, **2**, 1719.

Rowley, G., Saad, S., Giannelli, F., and Green, P. M. (1995). Ultrarapid mutation detection by multiplex, solid-phase chemical cleavage. *Genomics*, **30**, 574.

Sarkar, G., Yoon, H.-P., and Sommer, S. S. (1992). Screening for mutations by RNA single-strand conformation polymorphism (rSSCP): comparison with DNA-SSCP. *Nucleic Acids Research*, **20**, 871.

Sheffield, V. C., Cox, D. R., Lerman, L. S., and Myers, R. M. (1989). Attachment of a 40 base-pair G+C-rich sequence (GC clamp) to genomic DNA fragments by the polymerase chain reaction results in improved detection of single base changes. *Proceedings of the National Academy of Sciences, USA*, **86**, 232.

Sheffield, V. C., Beck, J. S., Kwitek, A. E., Sandstrom, D. W., and Stone, E. M. (1993) The sensitivity of single-strand conformation polymorphism analysis for the detection of single base substitutions. *Genomics*, **16**, 235.

van der Luijt, R., Khan, P. M., Vasen, H., van Leeuwen, C., Tops, C., Roest, P., et al. (1994). Rapid detection of translation-terminating mutations at the adenomatous polyposis coli (APC) gene by direct protein truncation test. *Genomics*, **20**, 1.

Verpy, E., Biasotto, M., Meo, T., and Tosi, M. (1994). Efficient detection of point mutations on color-coded strands of target DNA. *Proceedings of the National Academy of Sciences, USA*, **91**, 1873.

Wartell, R. M., Hosseini, S. H., and Moran, C. P. Jr (1990). Detecting base pair substitutions in DNA fragment by temperature-gradient gel electrophoresis. *Nucleic Acids Research*, **18**, 2699.

White, M. E., Carvalho, M., Derse, D., O'Brien, S. J., and Dean, M. (1992). Detecting single base substitutions as heteroduplex polymorphisms. *Genomics*, **12**, 301.

Winter, E., Yamamoto, F., Almoguera, C., and Perucho, M. (1985). A method to detect and characterise point mutations in transcribed genes: Amplification and over-expression of the mutant c-Ki-ras allele in human tumor cells. *Proceedings of the National Academy of Sciences, USA*, **82**, 7575.

6 The future

Mutation scanning can be divided into two categories: The detection of mutations in candidate disease genes and the identification of mutations in known disease genes, each of which have different requirements.

Towards the end of the twentieth century and the beginning of the twenty-first century, the detection of mutations in candidate disease genes will probably be based on the mapping of a particular phenotype to a particular chromosomal region, and the examination of all genes mapping to this region for mutations in order to identify the gene responsible. This will require a mutation detection method capable of screening large fragments of DNA, preferably with the ability to differentiate neutral polymorphisms from mutations.

The routine mutation screening of known disease genes will continue to be necessary to help understand the function of that gene and to assist the medical management of the patients concerned. This will require an automated method which either characterizes the mutation as well as detects it, or analyses fragments of a size which can be characterized in a single sequencing run. The key to both categories will be speed. Expensive equipment will be adopted if it automates the procedure (i.e. replaces you, and saves on your salary) and helps prevent human error.

As yet, no method is ideal for the detection of unknown mutations. SSCP is easy but does not detect all mutations and fragments analysed must be small. DGGE requires expensive PCR primers and the fragment length is usually dictated by the melting profile. CCM requires a high level of expertise and relies on the use of toxic and dangerous compounds. As the DNA sequence databases of mapped cDNAs increase, mutation detection becomes more of a rate-limiting step in the assignment of function to these genes. The development of new superior mutation detection methods is therefore one of the most important areas of molecular genetics.

It is impossible to predict the nature of future methodologies. A brief description of some novel methods currently under development is presented here.

1. Protein-dependent mutation detection

Any protein capable of binding to a mismatch in a DNA fragment can potentially be exploited for mutation detection. Protein binding could for example be recognized by an antibody, and the entire assay conducted by ELISA, without the need for gel electrophoresis. Some of the proteins capable of binding to DNA mismatches are enzymes which may be involved in the processes of DNA repair or recombination.

1.1 Mismatch cleavage by resolvases

The resolvases are a group of enzymes which catalyse the resolution of branched DNA intermediates formed during genetic recombination. These enzymes recognize such structures on the basis of bends or distortions in the DNA helix. The bacteriophage resolvases T4 Endonuclease VII (T4E7) and T7 Endonuclease I (T7E1), which are important in the handling of phage DNA during packaging and infection, respectively, are well characterized and have been purified. As well as being able to cleave synthetic Holliday (four-way) junctions, these enzymes can also cleave at or near single base-pair mismatches.

The technique is similar in basis to the other cleavage techniques (RNase and CCM), but involves fewer steps and avoids the use of toxic substances. In initial experiments, PCR-amplified fragments varying in size from about 100 bp to 1400 bp containing known mutations were heteroduplexed to ensure the presence of mismatches and subjected to enzyme cleavage analysis (Mashal *et al.* 1995; Youil *et al.* 1995). Approximately 90 per cent of mutations were identified. Some which were not recognized by T4E7 were recognized by T7E1 and vice versa, giving a combined detection rate of about 95 per cent. Unfortunately, the use of both enzymes together appears to be detrimental rather than beneficial. There is evidence that cleavage by these enzymes is dependent on both sequence context and DNA structure. Some cleavages are very weak and some are very strong, there being no consistency or correlation with the type of mutation. There is also a significant level of nonspecific cleavage, making the gels very 'dirty' in appearance.

More recent work has got the detection rate of T4E7 close or even up to 100 per cent (Youil *et al.* 1996). The use of magnetic bead technology to separate labelled homoduplex DNA away from the heteroduplex DNA has greatly improved the signal-to-noise ratio (Babon *et al.* 1995). The reduced background makes it much easier to spot specific cleavage products and reduces the possibility of a specific cleavage product being hidden behind a background band.

In many ways, resolvase technology promises to become the ideal method for mutation detection. But given that the attention of many researchers in this field is directed at developing a gel-free mutation detection system that is amenable to automation, the technique might be outdated before it has been perfected. In the immediate future, however (provided batch-to-batch enzyme consistency can be achieved), the efficacy and simplicity of the method give it a good chance of equalling or exceeding the popularity of SSCP and DGGE.

1.2 Mismatch recognition by DNA repair enzymes

DNA repair is another system which has the potential for exploitation in mutation detection. The *E. coli* mismatch correction systems are the best understood and as such are those on which the most attention has been directed. Three of the proteins required for the methyl-directed

DNA repair pathway, MutS, MutL, and MutH, are sufficient to recognize seven of the possible eight single base-pair mismatches (not C/C mismatches) and nick the DNA at the nearest GATC sequence. The MutY protein, which is involved in a different repair system, can be used to detect A/G and A/C mismatches (Lu and Hsu 1992). Some mammalian enzymes are also being investigated: thymidine glycosylase can recognize all types of T mismatch and 'all-type endonuclease' (Topoisomerase I) is capable of detecting all eight mismatches, but does so with varying efficiencies, depending on both the type of mismatch and the neighbouring sequence. When the MutY gene product, mammalian thymidine glycosylase and mammalian Topoisomerase I are combined, all types of mismatch can readily be detected. Furthermore the mutation can be characterized on the basis of which enzymes were able to detect it. Efforts are being made to streamline this approach, but is unlikely that it will ever offer any great advantages over any of the established methods of mutation scanning.

1.2.1 MutS

The MutS gene product is the methyl-directed repair protein which binds to the mismatch. Purified MutS protein has been used for mutation detection in several ways. Gel mobility assays can be performed in which DNA bound to the MutS protein migrates more slowly through an acrylamide gel than free DNA. This system has worked well for the ΔF508 cystic fibrosis mutation and has also been used to detect single base mismatches (Ellis *et al.* 1994; Lishansky *et al.* 1994).

An alternative version of MutS mismatch recognition, which does not require gel electrophoresis, involves the immobilization of MutS protein on nitrocellulose membranes (Wagner *et al.* 1995). Labelled heteroduplexed DNA is used to probe the membrane in a dot-blot format. When both DNA strands are used, all mismatches can be recognized by binding of the DNA to the protein attached to the membrane. Although C/C mismatches are not detected, the corresponding G/G mismatch derived from the other strand is recognized. The technique is very attractive in every way, it being simple, cheap, and amenable to automation. The detection efficiency, however, may be limited by the size of the DNA fragment. Although the system works excellently for short fragments (≤200 bp), there are doubts about its ability to handle larger fragments.

1.2.2 *In-vivo* use of repair enzymes

An alternative approach to using purified DNA repair enzymes in mutation detection is to use them in an *in-vivo* system. One such method has been devised (Faham and Cox 1995), in which hemimethylated heteroduplexed recombinant plasmids (containing the DNA fragment of interest) are transformed into *E. coli*. The plasmid also carries a *LacZα* reporter gene, which contains a 5 bp insertion on the methylated strand. If a mismatch is present in the test fragment, the *LacZ* insertion will be corrected, causing the bacterial colony to be white in colour rather than blue. Unfortunately, the entire procedure is very long and involves many steps. Despite the effort required, the system is still worthy of attention as it is capable of analysing fragments up to 10 kb in size

1.3 Structure-specific nuclease analysis of single-stranded DNA

◇ The CFLP system, developed by Third Wave Technologies, Madison, USA is marketed by Boehringer Mannheim and Life Technologies.

A thermostable, structure-specific nuclease termed 'Cleavase I' is being marketed for polymorphism and mutation detection. The system, known as 'cleavase fragment length polymorphism' (CFLP) technology, has much in common with both SSCP and the mismatch cleavage methods. It is based upon the production of folded, partially duplexed single-stranded DNA and the use of the cleavase enzyme to cut the DNA at the 5′ base of stem-loop structures. As the folded single-stranded structure varies between DNA molecules differing by as little as a single base pair, the Cleavase system is capable of detecting single base mutations.

The system is simple in that few steps are involved: The DNA is heated to 95 °C to denature, rapidly cooled to 55–60 °C at which point the cleavase enzyme and reaction components are added. The reaction is incubated at room temperature or above for just 1–5 minutes, before being stopped with a stop solution. Prior to electrophoresis on a denaturing polyacrylamide gel, the samples are heated to 80 °C for 2 minutes. Many bands are produced, giving a gel image similar in appearance to a DNA fingerprint or BarCode. Mutations are detected as the gain or loss of bands. Apart from its simplicity, the method is attractive in that it may be capable of analysing fragments significantly greater than 1 kb in size. It is unclear at this stage though, whether this system can be used to detect a high proportion of mutations without too much effort.

2. Sequencing by hybridization

The advent of fluorescence-based direct sequencing has made a huge impact on DNA sequencing. It has become much faster, is cheaper per base, and of course is more pleasant for the user as radiolabels are avoided. The speed of DNA sequencing may yet increase with the adoption of ultra-thin gels and capillaries. These gel systems dissipate heat much better than ordinary gels and can therefore accommodate much higher voltages. The basic principles of DNA sequencing however, have gone unchanged since the development of the chain-termination method by Fred Sanger and colleagues in the mid 1970s. In order for DNA sequencing to be truly revolutionized, fundamentally new methodologies need to be devised. One such hope for the future, is the concept of 'sequencing by hybridization' (SBH). In this method, arrays of short (8–10 base long) oligonucleotides are immobilized on a solid support in a manner similar to the reverse dot-blot (see Chapter 4, Section 1.3.1) and probed with a target DNA fragment.

The system is based on advanced chemistry in which the oligonucleotides are not synthesized separately and then fixed onto the support, but are synthesized together directly on the support (Fodor *et al.* 1993; Pease *et al.* 1994). The synthesis system begins with a glass chip coated with a nucleotide, linked to a light-sensitive chemical group which blocks the addition of further nucleotides. Specifically directed light is used to illuminate particular grid co-ordinates removing the blocking group at these positions. The chip is then exposed to the next photoprotected nucleotide, which polymerizes onto the exposed nucleotides.

In this manner, with successive rounds of nucleotide additions, oligonucleotides of different sequences can be synthesized at different positions on the solid support. Thirty-two cycles of specific additions (i.e. eight additions of each of the four nucleotides) should enable the production of all 65, 536 possible eight-mer oligonucleotides at defined positions on the chip.

When the chip is probed with a DNA molecule, e.g. a fluorescently labelled PCR product, fully matched hybrids should give a high intensity of fluorescence and hybrids with one or more mismatches should give substantially less intense fluorescence. The combination of the position and intensity of the signals on the chip should enable computers to derive the sequence. Thus with no need for electrophoresis and in a single hybridization, it might be possible to derive the sequence of any DNA fragment. In such a way, mutations could be diagnosed and characterized by a subtle difference in the fluorescence of the annealing pattern as compared with the wild-type.

This method is considered to many, to be the ultimate form of DNA sequencing and mutation detection. This dream however has problems, the most significant being that of internal annealing of the target molecule. The system requires that all parts of the fragment are annealed under a common hybridization condition. Some parts of the target fragment may contain regions of complementarity, causing intramolecular self-annealing. Any such region will not be available to hybridize to the oligonucleotide array, and so the region will be omitted from the deduced sequence. If this problem can be solved, it is possible that sequencing by hybridization will make an impact on mutation detection before it does so on DNA sequencing. For example, it should be possible to use the method to screen for all the known cystic fibrosis mutations before it is capable of sequencing DNA of unknown sequence.

Another hybridization strategy for the detection of known mutations is the measurement of hybridization kinetics of fully matched oligonucleotides versus mismatched oligonucleotides. The system consists of a biosensor, e.g. BIAcore (Pharmacia Biosensor), which monitors biological events in real-time by analysing surface plasmon resonance, enabling the detection of changes in refractive index at a sensor surface (Nilsson *et al.* 1995). These changes are proportional to the mass of molecules bound to the surface. From the results, stoichiometric and

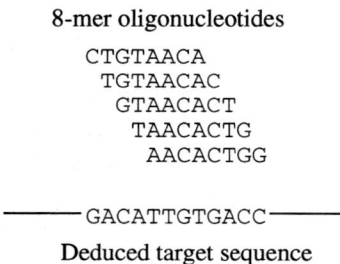

8-mer oligonucleotides

CTGTAACA
TGTAACAC
GTAACACT
TAACACTG
AACACTGG

———— GACATTGTGACC ————

Deduced target sequence

Fig 6.1

The target sequence is deduced from the overlapping sequence of the oligonucleotides with greatest fluorescence.

kinetic data for the interaction can be determined. The system can be used for the detection of protein–protein and DNA–protein interactions as well as DNA–DNA hybridizations. Mutation detection using such machines is in the early stages of development but success is a definite possibility. Unfortunately, the machines are extremely expensive.

3. Mutation detection by functional assays

It is possible to screen certain genes for mutations by introducing the gene into a cell system and testing for normal protein function. This approach has two fundamental advantages: large genes can be tested, as a single large DNA fragment is used, and polymorphisms go undetected.

Different organisms and cell types have advantages and disadvantages to offer every gene. However, the yeast, *S. cerevisiae*, shows the most promise for becoming the universal organism for functional assays. The main advantages of *S. cerevisiae* are the high level of biological understanding and its highly efficient system of homologous recombination. The latter allows foreign linear DNA fragments (e.g. PCR products) to be cloned into expression vectors, simply by co-transfection with linearized vectors.

If the vector is linearized and the ends carry regions of homology with the end sequences of the PCR product, then homologous recombination will efficiently 'clone' the PCR product into the vector. The use of centromere-bearing vectors allows the production of yeast colonies in which all the contained recombinant vectors are derived from a single homologous recombination event, i.e. contain a single foreign allele. Thus when a heterozygous PCR product is transformed, 50 per cent of colonies will contain one allele and 50 per cent the other.

The system depends upon the gene product concerned to have an activity which can be assayed on its ability to transactivate a reporter gene either directly by DNA-binding, or indirectly via interaction with another protein. The system has been used to screen the p53 gene for mutations on the basis of the DNA-binding activity of its protein product (Ishioka *et al.* 1993). The presence of p53 target sites upstream of the *HIS3* reporter gene was used to assay for normal or failed p53 function in yeast auxotrophic for histidine production. Expression of *HIS3* and hence the ability to grow on medium lacking histidine, signified normal p53 function. Failure of colonies to grow indicated p53 mutation. Reporter genes which cause the colonies to be one colour when expressed, and another colour when not expressed, offer an alternative to the use of auxotrophic mutants.

The system does have limitations, however. Proteins with more than one functional domain can only be tested for one domain at a time. Not all proteins can be assayed in this manner, e.g. structural proteins or proteins with no known interaction. It is also necessary that the cell system does not have any native proteins which can compensate for the loss of function of the foreign mutant protein. This is probably the most limiting aspect of the functional assay, but in time the use of yeast strains with such proteins removed may alleviate this problem.

To be worth the effort, the yeast functional assay must be used for testing large fragments. Given that these must consist of only coding sequences, the best approach to generating the fragment for analysis is by RT-PCR from RNA. This either requires access to the cells or tissues from the mutant individual in which the gene in question is expressed, or the amplification of 'leaky transcripts' (see Chapter 2, Section 1.2), which may require two rounds of amplification. The PCR itself requires a high fidelity polymerase such as *Pfu* (Stratagene) in order to minimize the number of incorporation errors.

The function of some genes can be directly assayed in patient-derived cell cultures. But a problem with this approach occurs when mutations are expected to be heterozygous, as the wild-type allele can mask the mutant allele. The problem can be solved by producing hybrid cells which contain only one allele (Papadopoulos *et al.* 1995), but this is a difficult and long-winded solution.

4. Denaturing high performance liquid chromatography

Reverse-phase high performance liquid chromatography is a method commonly used in the analysis of proteins. The technique allows the separation of proteins on the basis of the extent of hydrophobic interaction with a column matrix. In adapting the system to the detection of nucleic acid variants, a denaturing column is used which can discriminate between DNA homoduplexes and heteroduplexes, the heteroduplexes being released from the column before the homoduplexes (Jin *et al.* 1995). The eluate is examined by UV absorbance monitoring, detecting mutations as early-released DNA. The idea shows great promise as there is scope for automation and high throughput.

Further reading

Lipshutz, R. J. and Fodor, S. P. A. (1994). Advanced DNA-sequencing technologies. *Current Opinion in Structural Biology*, **4**, 376. Describes alternative DNA sequencing methodologies.

Modrich, P. (1991). The mechanisms and biological effects of mismatch repair. *Annual Review of Genetics*, **25**, 229. A thorough account of mismatch repair in *E. coli*.

References

Babon, J. J., Youil, R., and Cotton, R. G. H. (1995). Improved strategy for mutation detection—a modification to the enzyme mismatch cleavage method. *Nucleic Acids Research*, **23**, 5082.

Ellis, L. A., Taylor, G. R., Banks, R., and Baumberg, S. (1994). MutS binding protects heteroduplex DNA from exonuclease digestion *in vitro*: a simple method for detecting mutations. *Nucleic Acids Research*, **22**, 2710.

Faham, M. and Cox, D. R. (1995). A novel *in vivo* method to detect DNA sequence variation. *Genome Research*, **5**, 474.

Fodor, S. P. A., Rava, R. P., Huang, X. H. C., Pease, A. C., Holmes, C. P., and Adams, C. L. (1993). Multiplexed biochemical assays with biological chips. *Nature*, **364**, 555.

Ishioka, C., Frebourg, T., Yan, Y.-X., Vidal, M., Friend, S. H., Schmidt, S. *et al.* (1993). Screening patients for heterozygous p53 mutations using a functional assay in yeast. *Nature Genetics*, **5**, 124.

Jin, L., Underhill, P. A., Oefner, P. J., and Cavalli-Sforza, L. L. (1995). Systematic search for polymorphisms in the human genome using denaturing high-performance liquid chromatography (DHPLC). *American Journal of Human Genetics*, **57**, A124.

Lishansky, A., Ostrander, E. A., and Rine, J. (1994). Mutation detection by mismatch binding protein, MutS, in amplified DNA: Application to the cystic fibrosis gene. *Proceedings of the National Academy of Sciences, USA*, **91**, 2674.

Lu, A.-L. and Hsu, I.-C. (1992). Detection of single DNA base mutations with mismatch repair enzymes. *Genomics*, **14**, 249.

Mashal, R. D., Koontz, J., and Sklar, J. (1995). Detection of mutations by cleavage of DNA heteroduplexes with bacteriophage resolvases. *Nature Genetics*, **9**, 177.

Nilsson, P., Persson, B., Uhlen, M., and Nygren, P. A. (1995). Real-time monitoring of DNA manipulations using biosensor technology. *Analytical Biochemistry*, **224**, 400.

Papadopoulos, N., Leach, F. S., Kinzler, K. W., and Vogelstein, B. (1995). Monoallelic mutation analysis (MAMA) for identifying germline mutations. *Nature Genetics*, **11**, 99.

Pease, A. C., Solas, D., Sullivan, E. J., Cronin, M. T., Holmes, C. P., and Fodor, S. P. A. (1994). Light-generated oligonucleotide arrays for rapid DNA-sequence analysis. *Proceedings of the National Academy of Sciences, USA*, **91**, 5022.

Wagner, R., Debbie, P., and Radman, M. (1995). Mutation detection using immobilized mismatch binding protein (MutS). *Nucleic Acids Research*, **23**, 3944.

Youil, R., Kemper, B. W., and Cotton, R. G. H. (1995). Screening for mutations by enzyme mismatch cleavage with T4 endonuclease VII. *Proceedings of the National Academy of Sciences, USA*, **92**, 87.

Youil, R., Kemper, E. W., and Cotton, R. G. H. (1996). Detection of 81 of 81 known mouse β-globin promoter mutations with T4 endonuclease VII—the EMC method. *Genomics*, **32**, 431.

Glossary

Acrylamide One of the chemical monomers used to produce polyacrylamide gel.

Agarose A polymer extracted from seaweed used for electrophoretic separation of DNA molecules.

Allele A variant form of a gene.

Ammonium persulphate The catalyst for polymerization of an acrylamide gel.

Autoradiography A method for detecting radioactively labelled molecules using photographic film.

Biotin A molecule used as a nonradioactive label for DNA.

Bis See *N,N'*-methylenebisacrylamide

Bromophenol blue A marker dye commonly used in polyacrylamide gel electrophoresis.

Chemiclamp A psoralen-based linkage between the two DNA strands of a double-stranded DNA molecule. Sometimes used in denaturing gradient gel electrophoresis.

Chemiluminescence The chemical production of light, used in non-radioactive detection systems.

Chromogenic detection A labelling system which produces a visible precipitate.

Clone A population of genetically identical cells. May contain foreign DNA.

Cycle sequencing Successive rounds of DNA sequencing reactions, commonly performed in fluorescence-based sequencing.

Dideoxynucleotide A modified nucleotide that lacks the 3' hydroxyl group and so prevents further strand synthesis following its incorporation.

DNA polymerase An enzyme which synthesizes a new strand of DNA by polymerization of nucleotide subunits.

Dot blotter An apparatus for applying nucleic acid samples to a restricted area of membrane. Usually as small circles (or rods).

End-labelling Attachment of a marker to the 3' or 5' end of a DNA strand.

Ethanol precipitation Precipitation of nucleic acids by addition of ethanol and salt. Used to concentrate nucleic acids and to remove other substances.

Fixing The binding of DNA or protein to a polyacrylamide gel. Also washes urea out of denaturing polyacrylamide gels.

Fluorescence A detection system for the nonradioactive labelling of molecules.

Formamide A chemical denaturant commonly used in gels and gel loading buffers.

GC-clamp GC-rich DNA sequence attached to DNA fragments to create a region of high melting temperature. Frequently used in denaturing gradient gel electrophoresis.

Gel electrophoresis The size separation of molecules through a gel matrix using an electric current.

Gene A defined section of a DNA molecule which encodes a protein or functional RNA molecule.

Heteroduplex A double-stranded nucleic acid molecule in which the two strands are derived from different origins and are not completely identical in sequence.

Heterozygous The allelic state for a particular gene or site within a gene, in which the two alleles are of nonidentical DNA sequence.

Homoduplex A double-stranded nucleic acid molecule in which the two strands are identical in sequence.

Homozygous The allelic state for a particular gene or site within a gene, in which the two alleles are of identical DNA sequence.

Hybridization The formation of double-stranded nucleic acid by the base-pairing of complementary single strands.

Hybridization stringency The sequence specificity of nucleic acid double strand annealing. High stringency hybridization determines a high level of sequence complementarity. Low stringency hybridization permits a relatively high level of sequence non-complementarity.

Hydroxylamine A chemical which reacts with mismatched cytosine bases.

Karyotype The classified chromosomal content of an individual organism or cell.

Leaky transcription The very low level transcription which occurs when a gene is not actively expressed.

Ligase An enzyme which joins the 5′ phosphate and 3′ hydroxyl ends of nucleic acid molecules.

M13 primer An oligonucleotide originally designed as a DNA synthesis primer for sequencing DNA fragments cloned into the *E. coli* bacteriophage M13.

Melting temperature The temperature at which the two strands of a DNA molecule dissociate.

Mutation A change in the genetic material.

N,N′-**methylenebisacrylamide** Cross-linking component used in the production of polyacrylamide gels.

N,N,N′,N′-**tetramethylethylenediamine** The initiator of polymerization of an acrylamide gel.

Northern blot A technique for the transfer of RNA from an agarose gel to a membrane.

Oligonucleotide A short synthetic piece of DNA. Usually single stranded.

Osmium tetroxide A chemical which reacts with mismatched thymine bases.

Phenol extraction A method used to remove protein from nucleic acid preparations.

Piperidine A chemical which cleaves DNA at modified sites.

Polymerase chain reaction A method for the amplification of a particular region of DNA using a thermostable DNA polymerase.

Primer A short single-stranded oligonucleotide which acts as a starting point for nucleic acid synthesis.

Probe A labelled molecule used for the identification of another molecule. With nucleic acids, the probe is usually complementary in sequence to the target molecule.

Restriction endonuclease An enzyme which recognizes a specific double-stranded DNA sequence and catalyses the cleavage of the DNA strands.

Ribonuclease A group of enzymes which degrade RNA.

Southern blot A technique for the transfer of DNA from agarose gels to a membrane.

Streptavidin A molecule, often covalently attached to a solid surface, which binds with high affinity to biotin.

Taq **DNA polymerase** A thermostable DNA polymerase derived from the bacterium *Thermus aquaticus*.

TEMED See *N,N,N′,N′*-tetramethylethylenediamine

T4 polynucleotide kinase An enzyme used to attach phosphate groups to the 5′ ends of DNA molecules. Frequently used for end-labelling.

T7 RNA polymerase An enzyme which catalyses the synthesis of RNA molecules, derived from the bacteriophage T7.

Urea A chemical denaturant used in polyacrylamide gel electrophoresis.

Xylene cyanol A marker dye commonly used in polyacrylamide gel electrophoresis.

Index